A CULTURAL HISTORY OF MATHEMATICS

VOLUME 2

A Cultural History of Mathematics
General Editors: David E. Rowe and Joseph W. Dauben

Volume 1
A Cultural History of Mathematics in Antiquity
Edited by Michael N. Fried

Volume 2
A Cultural History of Mathematics in the Medieval Age
Edited by Joseph W. Dauben, Clemency Montelle, and Kim Plofker

Volume 3
A Cultural History of Mathematics in the Early Modern Age
Edited by Jeanne Peiffer and Volker R. Remmert

Volume 4
A Cultural History of Mathematics in the Eighteenth Century
Edited by Maarten Bullynck

Volume 5
A Cultural History of Mathematics in the Nineteenth Century
Edited by Tom Archibald and David E. Rowe

Volume 6
A Cultural History of Mathematics in the Modern Age
Edited by Tom Archibald and David E. Rowe

A CULTURAL HISTORY OF MATHEMATICS

IN THE MEDIEVAL AGE

VOLUME 2

Edited by Joseph W. Dauben, Clemency Montelle, and Kim Plofker

BLOOMSBURY ACADEMIC
LONDON • NEW YORK • OXFORD • NEW DELHI • SYDNEY

BLOOMSBURY ACADEMIC
Bloomsbury Publishing Plc
50 Bedford Square, London, WC1B 3DP, UK
1385 Broadway, New York, NY 10018, USA
29 Earlsfort Terrace, Dublin 2, Ireland

BLOOMSBURY, BLOOMSBURY ACADEMIC and the Diana logo are trademarks of Bloomsbury Publishing Plc

First published in Great Britain 2024

Series design by Raven Design
Cover image: Taqī al-Dīn and other astronomers at the Istanbul observatory of Sultan Murād III in the 1570s, as depicted in Seyyid Lokman's Persian *Book of King of Kings* (*c.*1581).
© Art Images via Getty Images

A catalogue record for this book is available from the British Library.

A catalog record for this book is available from the Library of Congress.

ISBN: HB: 978-13500-6294-8
ePUB: 978-13502-9479-0
ePDF: 978-13502-9478-3
Set: 978-13500-6312-9

Series: The Cultural Histories Series

Typeset by RefineCatch Limited, Bungay, Suffolk
Printed and bound in Great Britain

CONTENTS

ILLUSTRATIONS

Introduction

Chapter 3

Chapter 4

Chapter 5

Chapter 6

Chapter 7

CONTRIBUTORS

Sonja Brentjes is Visiting Scholar at the Max Planck Institute for the History of Science, Berlin, Germany. Her publications include *Teaching and Learning the Sciences in Islamicate Societies (800–1700)* (2018); co-edited with Sonja Brentjes, Jens Høyrup, and Bruce O'Brien, *Translation across Eurasia and Africa: From Babylonia to Colonial India* (2022); and edited with Peter Barker and Rana Brentjes, *Routledge Handbook on the Sciences in Islamicate Societies: Practices from the 2nd/8th to the 13th/19th Centuries* (2023). She is the recipientt of the Kenneth O. May Prize in the History of Mathematics, 2021, and the Research Award of the Annemarie Schimmel Foundation for Islamic Studies, 2022.

Jian-Ping Jeff Chen is a mathematician working in the field of the history of mathematics, specializing in the development of the subject after the Jesuits' introduction of European mathematics and astronomy into China in the seventeenth century. He received his BA from National Tsing-Hua University in Taiwan, and his MS and PhD in mathematics from Yale University, New Haven, CT, USA. Having worked on the history of trigonometry and "linear algebra" in China, his current research projects include the investigation of mathematical reasoning in late imperial China and the translation of the *Mathematical Arts in the Nine Chapters* and its commentaries into English, the latter of which for the first time incorporates the commentaries by certain scholars of the Song dynasty (960–1279).

Joseph W. Dauben is Distinguished Professor of History and History of Science at the City University of New York, USA. He is the author of *Georg Cantor, His Mathematics and Philosophy of the Infinite* and *Abraham Robinson: The Creation of Nonstandard Analysis, a Personal and Mathematical Odyssey*

(2014). A graduate of Harvard University (PhD 1972), he has been editor of *Historia Mathematica* and a former chair of the International Commission on the History of Mathematics. Professor Dauben has been a fellow of the American Academy in Rome, a member of the Institute for Advanced Study, Princeton Unniversity, is a Life Member of Clare Hall, Cambridge University, an honorary member of the Institute for the History of Natural Sciences of the Chinese Academy of Sciences, and recipient of the American Mathematical Society's Albert Leon Whiteman Memorial Prize for History of Mathematics, 2012.

Clemency Montelle is Professor in the School of Mathematics and Statistics at the University of Canterbury, Christchurch, New Zealand. She has research interests in the mathematical history of several early cultures of inquiry, including Mesopotamia, Greece, India, and the Islamic Near East. She is the co-author of several books on early mathematical astronomy, including *Editing and Analysing Numerical Tables towards a Digital Information System for the History of Astral Sciences* (2021) with editors Benno van Dalen and Matthieu Husson; *The Sanskrit Astronomical Table Text Brahmatulyasāraṇī Numerical Tables in Textual Scholarship* (2021) with Anuj Misra and Kim Plofker; *Sanskrit Astronomical Tables* (2019) with Kim Plofker; and *Chasing Shadows: Mathematics, Astronomy, and the Early History of Eclipse Reckoning* (2011).

Kim Plofker is Associate Professor of Mathematics at Union College in Schenectady, New York. She received her doctorate from the Department of the History of Mathematics at Brown University, Providence, RI, USA. Her research focuses on the history of mathematics and astronomy in India and its connections with Islamic and early modern European science. Professor Plofker's books include *Mathematics in India* (2009); and with Anuj Misra and Clemency Montelle, *The Sanskrit Astronomical Table Text Brahmatulyasarin*i (2021). She delivered a Plenary Lecture at the 2010 International Congress of Mathematicians in Hyderabad, India, and was awarded the Brouwer Medal of the Royal Mathematical Society of the Netherlands in 2011.

Nathan Sidoli received his PhD from the University of Toronto and is currently Professor of the History and Philosophy of Science at Waseda University, Tokyo, Japan. His research focuses on the Greek mathematical sciences and their development in Arabic sources. His publications include *Thābit ibn Qurra's Restoration of Euclid's Data* (2018) with Y. Isahaya; and *The Spherics of Theodosios* (2023) with R. S. D. Thomas. He is currently co-editor of *Historia Mathematica* and *SCIAMVS*, an annual journal dedicated to studies of the sources of ancient and medieval mathematical sciences, from any culture and in any language.

SERIES PREFACE

Culture, history, and mathematics are words that seldom appear together. Yet, mathematical activities have played a pervasive role in diverse cultures throughout human history, as described in the present six volumes. These illustrate how individuals and societies have engaged with various manifestations of mathematical culture from antiquity to the present. Based on recent and ongoing scholarship, they provide a global overview of a highly diverse range of developments, though certainly nothing like a comprehensive picture. Editors and authors had to make difficult choices about which topics and cultures to include, and in doing so, they have sought to strike a balance between characteristic features of a period and particular individuals and events that had a deep influence or longer lasting impact.

Each volume addresses seven general themes in as many chapters. The first three chapters —Everyday Numeracy, Practice and Profession, and Inventing Mathematics—highlight the roles of those who used, taught, or created mathematics as it evolved over time. These chapters explore mathematics per se, whereas the three that follow—Mathematics and Worldviews, Describing and Understanding the World, Mathematics and Technological Change—consider the larger impact of mathematics on other dimensions of human culture. The final chapter—Representing Mathematics—is devoted to a broader understanding of mathematical ideas, activities, or artifacts in different cultures throughout the world. The periods covered are:

Volume 1: A Cultural History of Mathematics in Antiquity
(3000 BCE–500 CE)
Volume 2: A Cultural History of Mathematics in the Medieval Age
(500–1400)

Volume 3: A Cultural History of Mathematics in the Early Modern Age (1400–1687)
Volume 4: A Cultural History of Mathematics in the Eighteenth Century (1687–1800)
Volume 5: A Cultural History of Mathematics in the Nineteenth Century (1800–1914)
Volume 6: A Cultural History of Mathematics in the Modern Age (1914–2020)

General Editors: David E. Rowe and Joseph W. Dauben

Introduction

JOSEPH W. DAUBEN, CLEMENCY MONTELLE,
AND KIM PLOFKER

The detail in Figure 0.1 from the cover illustration for this volume reflects the crucial and complex roles of mathematics in medieval societies, including its centrality to the great empires of China, India, and the Islamic world. This miniature painting of scholars at work in an astronomical observatory appears in an Ottoman manuscript containing the *Book of King of Kings* composed in Persian verse by the court historian Seyyid Lokman in the late sixteenth century. The observatory in question had been recently built in Istanbul to the specifications of the Arab mathematician-astronomer Taqī al-Dīn under the patronage of the Ottoman sultan Murād III. The sultan saw the creation of the observatory as a way of exhibiting the power of his monarchy and its financial strength, as well as indulging his own interest in astronomy and astrology. Taqī al-Dīn meanwhile was primarily concerned with improving the results recorded in the astronomical tables or *zīj* of the fifteenth-century Timurid sultan Ulugh Beg.

Murād III provided Taqī al-Dīn with everything he needed for his work, from financing the construction of the observatory's buildings, to employing a group of colleagues recruited from other renowned mathematicians and astronomers of Istanbul, to enabling easy access to the many books he required. Lokman's chronicle includes a poem containing a detailed account of the observatory's inspiration and activities:

> . . . The cultivation of all the sciences is indispensable to the theologians.
> This is especially true of Mathematics,
> For wise people grasp the subtleties of knowledge through the help of this department of science.

FIGURE 0.1. Astronomers at work in the Constantinople observatory in the 1570s, from a miniature painting in the late sixteenth-century Persian *Book of King of Kings* of Seyyid Lokman. Photo by Art Images via Getty Images.

Accounting depends on its branch called arithmetic . . .
The determination of planetary positions, as well as the construction of astronomical tables,
Is made by those well-versed in that science. . . .
It is through this science that one comes to know with certainty
The astronomical tables of Ulugh Beg and Naṣīr al-Dīn al-Ṭūsī
Had become worn-out like traces of mats upon soft soil . . .
When he [Sultan Murād] issues orders for making observations and compiling astronomical tables,
The stars will descend and prostrate themselves before him,

And when, in their endeavors, the astronomers are backed up by his
sovereign power,
They will carry off the crown from the stars of the Ursa Minor.
Whosoever compiles the new tables during his reign
Will become freed from observation programs till doomsday.
(Sayili 1956: 471ff.)

In the miniature, several of the observatory's staff members (comprising observers, clerks, and assistants) are depicted using some of its mathematical instruments such as an astrolabe, a quadrant, various rulers and sighting devices, a set square, and a mechanical clock, some of which were invented or improved by Taqī al-Dīn himself. The shelves of colorfully bound manuscripts represent part of the institution's library of works on mathematics and astronomy, and the general impression is one of enthusiastic activity. Yet in 1580, just a few short years after its completion, the Istanbul observatory was destroyed by order of the sultan and its mission abandoned.

The fate of the Istanbul observatory can be read in several ways as illustrating widespread changes accompanying the end of the medieval world and the remaking of its mathematics. Contemporary sources recount that Taqī al-Dīn inferred from his observations of a comet in 1577 an auspicious omen for an Ottoman attack on their Safavid rivals—an omen which the subsequent course of the war did not fulfill. The exasperated sultan, responding to queries from "people of learning"—and, according to the poem, with the concurrence of Taqī al-Dīn—obliterated the offending institution. We return to this aspect of our story at the end of this introduction, where we follow the long second-millennium evolution of mathematical traditions across the Eurasian continent, yielding to a more monolithic modern global mathematics beginning in the nineteenth century.

But the main focus of our story is the remarkable diversity and growing power of medieval mathematics itself. Sometimes assimilated from foreign sources, sometimes expanding through original and independent developments, numerate learning in the fifth through fifteenth centuries flourished from China and India to the Islamicate countries of Central and Western Asia and Northern Africa, and onward to the Byzantine and Latin realms. Throughout, the currents of transmission continued to shift and multiply: from east to west and vice versa, from monks to merchants, from Muslims to Confucians (to name but a few examples), the circulation of mathematical ideas flowed on.

TEMPORAL AND GEOPOLITICAL FRAMEWORKS

In popular portrayals of (Western) European history, the "Middle Ages" or "Dark Ages" even nowadays are often described as a period of retreat and isolation, a turning away from the ancient cosmopolitan empires into ignorant dogmatism. This was never a particularly accurate picture of medieval Europe, and it is still less applicable to the rest of the world. Even the chronological boundaries of this period are at best an unsatisfactory compromise, derived from the traditional bracketing of the European "Middle Ages" by the extinction of the Western Roman emperors shortly before 500 CE and the Ottoman conquest of the Byzantine or Eastern Roman Empire in 1453.

We have interpreted the temporal limits of our period somewhat more generously in discussing mathematics in non-European regions, where our coverage will overlap to some extent with material in earlier and later volumes. In the setting of Chinese mathematics, for example, our discussion segues from ancient to medieval more or less in the Northern and Southern dynasties starting in 386, and continues as far as the Ming Dynasty that ended in 1644. In South Asia, we put the corresponding transition to the medieval period a century or so after the Satavahana and Kushan rulers in the early first millennium, and we follow subsequent developments in mathematics nearly up to the waning of the Mughal Empire after 1700. Over approximately the same time period elsewhere in Asia and North Africa, our purview extends more than a thousand years from the later Sasanian Empire in the mid-first millennium to the later Ottoman Empire. Roughly, then, our version of the "medieval" extends from the aftermath of the expansion of ancient empires to the launch of early modern European imperial enterprises in Africa, Asia, and the New World.

The standard geographical boundaries of the medieval world are dominated by Asia, stretching from the Pacific in the East to the Mediterranean in the West. However, positioning the modern continent of Asia geographically on a world map is as much a political and social construction as a matter of topography. Conventionally, the Red Sea and Suez Canal are taken to separate Asia from Africa, and the Bosporus and the Urals (both the river and mountains) constitute the boundaries between the northwestern parts of Asia and Europe.

The name "Asia" is first identified in literature by the fifth-century BCE Greek historian and geographer Herodotus. He considered the continent designations of Europe, Asia, and Libya (Africa), with rivers such as the Nile demarcating their boundaries, to be an essentially arbitrary division of a single unified landmass. The civilizations within that landmass that constitute the subject of this study were defined as much by their geography and their great rivers that sustained life and ensured prosperity as by the distinct languages and ethnicities of their native populations.

India and China were easily the wealthiest and most prosperous of regions in the world for much of the medieval period. China was surrounded by great mountain ranges like the Himalayas and vast deserts like the Taklamakan and Gobi deserts, while its two great waterways, the Yangtze and Yellow rivers, flowing for the most part from west to east, were essential to the earliest urban centers that depended upon them for the success of agriculture and trade alike. During the Sui and Tang dynasties in the latter half of the first millennium, the Grand Canal was completed, facilitating the transport of grains and goods between the prosperous agricultural south and the northern regions of China. The Tang capital Chang'an (modern-day Xi'an) was the world's most populous city, numbering as many as 80 million people. The population of China nearly tripled from some 50 million in the early Tang dynasty to 140 million or so in the early second-millennium Song, which was also the period of the greatest innovations in mathematics. In addition to the prosperity of agriculture, the silk and porcelain industries contributed substantially to the expanding economy and wealth of the country. So did technological advances that improved the quality and quantity of goods available both locally and for foreign export.

The lands known to their western neighbors by such names as "India" or "al-Hind," already renowned in China as the cradle of Buddhism, evolved their own imperial states mostly within the religious-cultural matrix of what has become known as "Hinduism." During the reigns of regional dynasties such as the Guptas, Pallavas, Rashtrakutas, and Cholas, inhabitants of the South Asian subcontinent constituted around half of the world's population. Indian production of commodities including textiles, indigo, sugar, and pepper, as well as re-exported Chinese goods, made the "gold of Inde" a byword in the medieval and early modern world. Not only Indian merchandise but also Hinduism and Buddhism, together with Classical Sanskrit literary and scientific scholarship, spread beyond the subcontinent to Southeast Asia.

Meanwhile, to the west, invading tribes from the north had begun by the end of the fourth century to destabilize Rome's European frontier. In 476, the barbarian Flavius Odoacer deposed the last Roman emperor, Romulus Augustulus, marking the end of Roman hegemony. The ancient rivalry of Rome and Persia was perpetuated by the empires of Eastern Rome or Byzantium, ruled from Constantinople (today's Istanbul), and the Sasanian Persians, succeeded in the mid-seventh century by the emergent Muslim caliphates. Sasanian systems of government administration and the Persian language remained important in the rapid growth of Islamic empires, particularly in their eastward expansion, while the Muslim faith and the Arabic language of its scripture became defining features of the "Islamicate" world. Eventually extending over much of the Eurasian and African continents from Spain to Southeast Asia, the territories of *Dār al-Islām* controlled many of the world's wealthiest trade routes and cash revenues. The Islamicized Mongol khanates and Turco-Persianate empires (including the Ottomans who overpowered their Byzantine rivals in 1453, and the Mughals in northern India) continued to dominate the lands between Europe and China until well into the modern period.

During this millennium Europe was largely a fragmented collection of feudal states and rival empires. On Christmas Day in the year 800, when the population of Europe is estimated to have been between 25 and 30 million, Pope Leo III crowned Charlemagne, formerly ruler of the Franks and Lombards, as king of what would later become known as the Holy Roman Empire. Although, as Voltaire famously remarked, his realm was neither holy, Roman, nor an empire, Charlemagne did succeed in uniting a large part of medieval Western Europe under one ruler for the first time. (The empire would persist under various names and reshufflings of its boundaries until it was finally dissolved early in the nineteenth century as a result of the Napoleonic wars.)

Medieval Europe was divided by the schism of the Western and Eastern Orthodox Churches in 1054, and further by the creation of the Papal States in the 1177 Treaty of Venice. By the end of the thirteenth century, thanks in part to a prolonged period of warmer temperatures and increased agricultural productivity now known as the Medieval Warm Period, the population of Europe grew to 70 million. But after a series of famine-producing crop failures and diseases among farm animals followed by a series of pandemics beginning with the Black Plague, the fourteenth-century population of Europe fell by 50 percent or more, and economies suffered commensurately. There was no concentration of power or wealth in Western Europe comparable to the empires of Asia in the medieval period; scholarship and science were mostly a matter for learned clerics of the Church and their monastic schools. But as royal authorities in the early second millennium permitted (or in some cases decreed) the creation of corporate bodies of instruction for guilds of scholars, many of whom knew something of classical and Arabic works, academic engagement with mathematical sciences increased.

Naturally, medieval people the world over shared with humans in other ages the fundamental mathematical tasks of quantifying time, space, and money. And scholars among them continued to speculate about new ways to develop mathematics. But most of what we know—and almost all of what we can discuss in this volume—about mathematics during the medieval "long millennium" comes from textual sources in the main literate traditions of the Afro-Eurasian landmass. These traditions are here somewhat arbitrarily identified with written works in Chinese, Sanskrit, Arabic, Persian, Greek, and Latin. This means that we unfortunately neglect a great deal of valuable information about contemporary mathematical developments outside these major literate knowledge systems in the northeastern or "Old World" quadrant of the globe. The base-20 astronomical calculations of the Maya in prehispanic Mesoamerica, the celestial navigation techniques of Polynesian seafarers, the additive/multiplicative numeral-word system of the Pomo people of North America, and combinatorial strategies in various forms of pit-and-counter mancala games in many African cultures are just a few examples of the many interesting expressions of medieval mathematical thought to which we cannot do justice here.

Also regrettably abbreviated in this cultural history are the innumerable connections between the literate traditions that we have taken as our major focus and some of their less well-known neighbors. Such mathematical interactions include those of Chinese works with Korean and Japanese texts, of Sanskrit works with south Indian Dravidian or northern Middle Indic ones, of Persian texts with Turkish ones, of Greek or Arabic scholarship with Syriac, and of Latin with European vernaculars such as German or Italian. We will treat mathematical sources in these smaller, more local linguistic corpora only occasionally, usually in the context of their exchanges with their more broadly influential counterparts.

Throughout the medieval world mathematical traditions were linked to one another by the wider political and economic relations of cultures and states, especially in the medieval expansion of international trade. Their connections permitted sustained engagement with other cultures' views of the world, both spiritually and materially. Besides the transcontinental spread of religious systems such as Buddhism, Christianity, and Islam, this period saw increasing diffusion of mathematical techniques related to circular and spherical models of the universe and systems of celestial divination derived from them. Laypeople and less elite practitioners also developed a common heritage of mathematics in the form of practical and recreational problems and ways of solving them, which became part of the core of elementary mathematics education.

But in every medieval culture the subject matter of mathematics also had deep roots in a culturally specific ancient body of learning. It is as impossible to understand the Chinese or Sanskrit mathematics of the cosmopolitan medieval world without knowing something of its own more local forebears as it would be to study, say, Thomas Bradwardine or Copernicus with no awareness of Euclid or Aristotle. Therefore we turn in the following section to a brief survey of some of the mathematics of ancient East and South Asia, as a counterpart to the roughly contemporaneous Mediterranean-region developments in Mesopotamia and Greece that were discussed in the preceding volume.

LEGACIES FROM ANCIENT CHINA AND INDIA

Ancient Chinese Mathematics

The origins of mathematics are attributed in various Chinese legends to several mythical figures. The second-century BCE historian Sima Qian recounts that 2,000 years previously Emperor Yü had sought to control perennially flooding rivers "with a plumb-line in his left hand and a gnomon and compass in his right." His legendary predecessor the Yellow Emperor assigned his subjects various tasks for observing the sun, moon, and stars, fixing musical scales, determining means for telling time and ordering the seasons, and creating arithmetic. All these narratives testify to a conception in ancient China of mathematics as a fundamental aspect of human civilization, whose basic tools and techniques had been established in remote antiquity.

Modern historiography takes up the story around 1800 BCE when the first walled cities began to appear in China, and with them the long history of increased warfare between rival kingdoms. The power of kings depended not only on the maintenance and provisioning of their armies, but on land taxes and levies payable by landowners or peasant farmers, usually in grain or other locally produced commodities. This all required mathematically skilled administrators who were responsible for the computation and collection of tribute money.

The first written records in China date from the Shang dynasty period in the late second millennium BCE, in the form of divinatory inscriptions on bones and turtle shells. Remarkably, there are two distinctly different numbering systems reflected in the numbers found on these "oracle bone" inscriptions. One was used for counting days in repeating two-month cycles of 60 days, based on other cycles of lengths 10 and 12. The other is a decimal but not strictly place-value system, combining specific characters for the digits 1 through 9 with additional symbols for the powers of ten:

1	2	3	4	5	6	7	8	9	10
甲	乙	丙	丁	戊	己	庚	辛	壬	癸

The Ten Heavenly Stems (*S*)

1	2	3	4	5	6	7	8	9	10	11	12
子	丑	寅	卯	辰	巳	午	未	申	酉	戌	亥

The Twelve Earthly Branches (*B*)

1 甲子	2 乙丑	3 丙寅	4 丁卯	5 戊辰	6 己巳	7 庚午	8 辛未	9 壬申	10 癸酉
11 甲戌	12 乙亥	13 丙子	14 丁丑	15 戊寅	16 己卯	17 庚辰	18 辛巳	19 壬午	20 癸未
21 甲申	22 乙酉	23 丙戌	24 丁亥	25 戊子	26 己丑	27 庚寅	28 辛卯	29 壬辰	30 癸巳
31 甲午	32 乙未	33 丙申	34 丁酉	35 戊戌	36 己亥	37 庚子	38 辛丑	39 壬寅	40 癸卯
41 甲辰	42 乙巳	43 丙午	44 丁未	45 戊申	46 己酉	47 庚戌	48 辛亥	49 壬子	50 癸丑
51 甲寅	52 乙卯	53 丙辰	54 丁巳	55 戊午	56 己未	57 庚申	58 辛酉	59 壬戌	60 癸亥

The Ten Heavenly Stems and Twelve Earthly Branches, used in combinations that uniquely identify each day in two-month cycles of sixty days.

To compute the number *N* for any given day in this system is simply a matter of knowing the stem number *S* and the branch number *B* of the day in question, from which it then follows that $N = 6S–5B$. There is a problem related to exactly this sexagesimal cycle in one of the earliest mathematical works from ancient China, the *Book on Numbers and Computations*, Problem 50, which states: "A person travels for 50 days; now, today is *ren shen* (壬申) day; on which day did the trip begin?" The answer is that the trip began on *ren wu* (壬午) day. This can be seen from the above table, because the *ren shen* day is a combination of the 9th of the Heavenly Stems and the 9th of the Earthly Branches, hence, $N = (6)(9)–(5)(9) = 54–45=9$. Likewise, the number for the *ren wu* day is $N = (6)(9)–(5)(7) = 54–35 =19$. To solve this problem, one has to subtract 50 from 9; this will mean subtracting 41 days from the previous 60 day cycle, or 50–41= 19, and that indeed gives the *ren wu* day on which the trip began.

This system of identifying days of the year in terms of sixty-day cycles was later extended to counting the lunar months, and from the second century BCE, years were also counted in the same way, a system that is still in operation today in terms of naming the years of the Chinese calendar. It should be noted, however, that this sexagesimal system was never used in computations or in applications to astronomy, unlike the ubiquitous use of the base-60 system in Mesopotamian mathematics and later Greco-Roman astronomy.

The second numerical system to be found in use among the oracle bone inscriptions is a decimal system, based upon symbols for units, multiples of ten and one hundred, with additional symbols for thousands and ten thousands. The largest number yet found among those on oracle bones is 30,000. This would eventually evolve into the system that is still used today to represent numbers in China, whose modern equivalents below evolved from earlier seal characters on the ancient bones (*c.* 1250 BCE) to bronzeware script, small seal script, and clerical (official) script of the Han dynasty:

1	2	3	4	5	6	7	8	9	10	100	1000	10.000
一	二	三	四	五	六	七	八	九	十	百	千	萬

Thus, to express a number like 12,254, this would be written down as 1 ten-thousands, 2 thousands, 2 hundreds, 5 tens, and 4: 一萬二千二百五十四; whereas a number like 10,204 would simply be expressed as 一萬二百四, i.e. 1 ten-thousands, 2 hundreds, and four. In this system there was no need for a zero because wherever it might occur in a given number, as in 10,204, the zero was effectively expressed by the absence of the symbol for a thousand or ten.

But when it came to manipulating rather than recording numbers, the Chinese adopted another system that involved counting rods that were used in carrying out actual mathematical computations. These were rods that, when placed down on a surface—henceforth called a counting board—could represent any needed number, but these rods could then be used to perform all of the basic arithmetic operations and were even used to find square and cube roots.

Eventually, as Chinese mathematics progressed, the rods could even serve to solve polynomial equations of arbitrary degree, equations in as many as four unknowns, and sets of simultaneous linear equations:

The rods, depending upon their orientation, could be displayed either vertically:

1	2	3	4	5	6	7	8	9
𝍠	𝍡	𝍢	𝍣	𝍤	𝍥	𝍦	𝍧	𝍨

or horizontally:

10	20	30	40	50	60	70	80	90
𝍩	𝍪	𝍫	𝍬	𝍭	𝍮	𝍯	𝍰	𝍱

This distinction was important to avoid mistakes in representing any given number on the counting board; the vertical rods were used to represent units, hundreds, ten-thousands, i.e. even powers of ten, whereas the horizontal rods were used to represent powers of ten, thousands, i.e. the odd powers of ten. This was a positional decimal system in that positions on the counting board from right to left represented units, tens, hundreds, etc. Thus, 23 would have been represented by 𝍪𝍢. Had 𝍡 been used in the ten's place to represent 20, this might have been mistakenly read as 5, i.e., 𝍡𝍢. In this way, alternating the horizontal and vertical placement of the rods, there was little likelihood that as they were laid out on the counting board, it would be unclear whether their value should be tens, hundreds, thousands, etc. Thus, in terms of the counting board configuration, the following would have represented 12,254: 𝍠𝍪𝍡𝍭𝍣. Again, there was no need for a special rod to represent zero, because leaving a position empty would signify that there were no tens or thousands, for example, 10,204 would have appeared as follows when represented by rod numerals: 𝍠 𝍡 𝍣.

By the Warring States period near the start of the Common Era, the quasi-place-value system of written numerals coexisted with the rod-numeral system used for manipulating numbers in mathematical computations rather than simply recording them. When aligned on a flat surface (counting board), these small bone or bamboo rods could be used not only to represent any desired number, but also to perform all the basic arithmetic operations including square and cube roots. Eventually, as Chinese mathematics developed, the rods even served to solve polynomial equations of arbitrary degree, equations in as many as four unknowns, and sets of simultaneous linear equations (for details, see Chapter 3).

Copies of Chinese texts systematically describing these computational practices survive from around this time. These compositions, including two called *Mathematics* and *Book on Numbers and Computations* which are written on bamboo strips, describe the basics of multiplication and division of numbers, how to deal with fractions, and various mathematical problems that would have occupied government bureaucrats in the course of their official duties involving the collection of taxes or the organization and payment for government projects like building walls and digging canals.

Mathematics was also important for setting the official dynastic calendars as well as for issuing annual almanacs upon which common people depended for determining auspicious days for planting and harvesting, for undertaking financial transactions or new business, or indicating inauspicious days when any important undertakings should be avoided. Numbers as they were used in the predictions of the divinatory classic *Book of Changes* also affected the lives of ordinary people who sought to determine their future courses of action based on its mystic hexagrams.

By the early years of the Han dynasty around 200 BCE, the basic body of mathematics had been established in two works that later became what are today deemed "mathematical classics," the *Mathematical Classic of the Zhou Gnomon*, a work on astronomy and cosmology discussed in greater detail in Chapter 5, and another, the famous *Nine Chapters on the Art of Mathematics*, whose major innovations signal the beginning of new mathematical results of the medieval period discussed in Chapter 3. These works further elaborated procedures for dealing with fractions, finding the areas and volumes of various geometric figures, and computing rates of exchange for various commodities used in the Chinese barter economy, including grain exchanges and the use of silk as well as coins in handling commercial transactions.

Ancient Indian Mathematics

The fragmentary record of early numeracy in South Asia is preserved in artifacts and building sites constructed between 3500 and 1500 BCE in the Harappan or Indus Valley cultures, and in some of the orally transmitted Old Indo-Aryan or Vedic Sanskrit sacred texts. No conclusively attested number words or numerals have been deciphered in Harappan glyphs, although metrological artifacts figure prominently in several Harappan archaeological sites.

The ancient sacred texts of the Vedas (from Sanskrit *veda*, "knowledge") were passed down among Sanskrit speakers by recitation and memorization from at least the second millennium BCE. The occasional number words they contain include not only names of familiar decimal numbers and fractional parts, but also terms for successive powers of ten well beyond the thousands and ten-thousands familiar to other ancient cultures—as far as about 10^{12}. The Vedas likewise allude to elements of ritual calendars using intercalary months to synchronize lunar and solar cycles. By the mid-first millennium BCE, all these mathematical subjects and many others had been expounded in various Sanskrit auxiliary texts of the Vedic corpus.

Most notably, the *Śulbasūtra* or "cord-rule" precepts described a system of ritual geometry involving complicated constructions of various figures for sacrificial altars of baked brick as well as area-preserving transformations from one figure to another. They attest to contemporary knowledge of mathematical concepts including the "Pythagorean" right-triangle relation, and specific numerical values for constants such as the ratio of a square's diagonal to its side and the ratio of a circle's circumference to its diameter. Other texts preserve computational rules for calendar maintenance and timekeeping by means of water clocks and gnomon shadows. It has been speculated that Mesopotamian mathematical astronomy may have influenced, via contact with the Achemenid Empire, some technical features of its Indian counterpart such as sexagesimal divisions of time-units, but no corroborating evidence of such transmission has been found.

In the last few centuries before the Common Era, a growing corpus of philosophical thought in the Upanishads of Brahmanism/Hinduism, as well as in Buddhist and Jain works, explored metaphysical concepts including ideas of zero and infinity. Philosophers' speculative cosmological models hypothesized vast extents of space and time constructed from aggregates of infinitely tiny units of measurement. Scholars in more mundane fields also left some textual traces of mathematical knowledge and development. For example, combinatorial techniques are attested both in medical encyclopedias (in the context of combinations of humoral imbalances and of principal flavors) and in a work on prosody (to find, e.g., numbers of possible combinations of heavy and light syllables). Quantitative techniques retained a significant role in several canonical Sanskrit disciplines or *śāstras*, but were overwhelmingly concentrated in the *śāstra* of astronomy/calendrics.

Meanwhile, increasingly urbanized societies around the Mauryan Empire established administrative bureaucracies heavily dependent on quantitative techniques, dealing with numerical data ranging from taxes and tariffs to government budget revenues from mines, forests, salt manufacture, gambling establishments, and many other industries. The social roles of mathematical practitioners among Hindus at this time were largely determined by the traditional four *varṇas* or social categories designated *brāhmaṇa*, *kṣatriya*, *vaiśya*, and *śūdra*, which were further subdivided over time into the complex array of kinship groups and hereditary occupations now generally described as "caste." The hereditary professions of religious officiants and scholars, including astronomer-astrologers, were occupied almost exclusively by *brāhmaṇa* groups. Scribes and administrators, especially at royal or noble courts, might be *brāhmaṇas* or *kṣatriyas* (traditionally rulers and warriors), but the scribal profession also evolved an almost independent social-occupational category called *kāyastha*. Traders and producers were assigned to the *vaiśya* category while laborers and artisans were *śūdras*. Individuals or groups might occasionally change their caste status due to a shift in wealth, prestige, or claimed identity. And all groups had their own traditions of literate or oral/craft quantitative knowledge, as did their contemporaries among Jains and Buddhists.

Most of the surviving texts from which we draw our knowledge of ancient and medieval literate Indian mathematics (and many more such works that have not survived) were composed by authors professing *brāhmaṇa* affiliation and astronomical/astrological expertise. These include the textual tradition of early Common Era Sanskrit astral science incorporating many of the fundamental concepts of Hellenistic spherical astronomy and horoscopic astrology known to post-Alexander Indo-Greek inhabitants of the northwestern part of the subcontinent. It appears that in the evolutionary process of this Sanskrit mathematical astronomy/astrology—perhaps influenced by trade contacts with the rod numerals of China and/or by the Greco-Babylonian sexagesimal place-value system—Indian calculators adumbrated the decimal place-value numbers eventually adopted worldwide, whose story is told in more detail in Chapter 1.

EXCHANGES AND TRANSFORMATIONS

The evolution of medieval mathematics across centuries and continents, and the gaps in its documentary "fossil record" that often can only be filled with guesswork, are neatly illustrated by a classic study of a practical-geometry technique known as the "double difference method" (Figure 0.2). This technique enables the user to determine without trigonometry the height and distance of some inaccessible object (for example, an island peak across a stretch of sea) by taking separate sightings of it from a known height at two locations a known distance apart. The two non-parallel rays from the sighting-points to the top of the distant object form right triangles that can be solved for the desired measurements.

The historian of mathematics Kurt Vogel noted nearly forty years ago in his article, "A surveying problem travels from China to Paris," that the double difference method is first documented in a supplement to the Chinese mathematical classic *Nine Chapters* composed in 263 CE by the commentator Liu Hui, which was later known as an independent work entitled *Sea Island Mathematical Manual*. Preserved in Chinese texts for many centuries, this technique is hypothesized to have made its way to India at some point prior to the composition of the Sanskrit treatise *Āryabhaṭīya* by the astronomer-mathematician Āryabhaṭa around 500. In Āryabhaṭa's version the distant object is a lamp at the top of a post, and its height and distance are derived from the lengths of the shadows cast by two vertical gnomons of known height. This

FIGURE 0.2. Illustration of the double-difference method for finding the height of a distant mountain, as described in *The Sea Island Mathematical Manual* by Liu Hui, originally composed in 263 CE; this image accompanies a version of the text published in 1726 (Needham 1959: 572). Courtesy of the Needham Research Institute, Cambridge, UK.

form of the double difference problem is retained in later Indian mathematical works such as that of Brahmagupta in the early seventh century, where it is treated as part of the standard mathematical topic called "shadows" (see Chapter 3).

The renowned eleventh-century Muslim scientist Abū Rayhān al-Bīrūnī, who lived for many years in northwest India and wrote extensively about Sanskrit texts and Indian cultures, was familiar with Brahmagupta's writings. His *Exhaustive Treatise on Shadows* explicitly mentions Brahmagupta in discussing the double difference method, a rare instance of clear documentary evidence of cross-linguistic transmission of medieval mathematics:

> Brahmagupta explained in the [above]mentioned book, thus, "If there is a lamp on a stand, with an obstacle intervening between us and its base, and near [us] is a gnomon whose length is twelve digits, and behind its shadow is another gnomon of that size, its shadow being eighteen digits, and from the head of the first shadow to the base of the second gnomon is seven digits, and one [wants] (to find) the length of the stand." (Kennedy 1976: 256)

Moreover, al-Bīrūnī apparently adopted these numbers from a worked example appearing in a ninth-century commentary on Brahmagupta's text. But in his presentation, Brahmagupta's problem is just a variant form of the main theme of his chapter, "On the determination of terrestrial distances and the heights of mountains by (the use of) shadows," for which his solution techniques generally involve sighting by means of an astrolabe.

Roughly contemporary with al-Bīrūnī's treatise are the earliest known manuscripts of a Latin work on practical geometry and surveying which also describes the double difference method, employing either vertical sighting-posts or astrolabe observations. Such problems of so-called two-station altimetry using the astrolabe were standard in late medieval examples of this genre, including two Parisian works both entitled *Practices of Geometry*: one from the twelfth century (attributed to the abbot Hugh of St. Victor) and one written in 1346 by Dominicus de Clavasio at the University of Paris.

What can we confidently deduce from these episodes concerning the historical origin and spread of the double difference method? It seems clear that it first became established in the textual tradition in post-Han dynasty China. From there it was likely brought to Gupta-era India, possibly by Chinese scholars on pilgrimage to Buddhist sites, although there is no direct evidence of transmission to rule out the possibility of independent development within Sanskrit geometry. Al-Bīrūnī testifies that the standard Indian form of the method was unquestionably known to some Islamic mathematicians before the twelfth century, but treated as peripheral to geometrically similar altimetry techniques using the Greco-Islamic astrolabe. Medieval Europeans certainly learned about the astrolabe and associated observational procedures involving surveying and trigonometry from Islamic sources. However, these procedures along with some form of the double difference method had apparently already entered Latin texts before any Latin scholar could have encountered al-Bīrūnī's references to Brahmagupta's problem about finding the height of a lamppost from two gnomon shadows.

In short, the history of the double difference method is not so much the story of a single intrepid wanderer trekking across a continent, but instead a complex multigenerational process of evolutionary dispersal. The same holds true for a great many of the documented or hypothesized large-scale mathematical exchanges in the medieval world. Ideas and techniques may spread slowly and locally or suddenly and far, in the process intermingling with other ideas and problems, creating new hybrids and species of mathematical knowledge finding new niches in their intellectual environment. Meanwhile, other forms of mathematical knowledge may appear to die out from cultures where they once flourished, only to be rediscovered or reinvented in some future time and place.

The remainder of this section samples some more directly attested transmissions of mathematical sciences, in both eastward and westward directions. For example, during the Tang dynasty in China a few court scholars of South Asian descent introduced to their colleagues some Indian works, including material on astronomy and astrology. One of them named Qutan Xida (Gautama Siddha or Siddhārtha) helped compose in the early eighth century what appears to be the first Chinese work to incorporate elements of Indian trigonometry and the Indian decimal numerals with a circular zero.

By the mid-eighth century, Islamic mathematical sciences had begun to emerge as a recognized field of knowledge conveyed mostly in Arabic and Persian, but radically influenced by foreign text corpora. The court at the new ʿAbbāsid capital of Baghdad, in conscious imitation of similar initiatives in the late Sasanian Empire, sponsored a sustained program of scholarly translation into Arabic. Translation efforts in exact sciences initially focused on Sanskrit and Pahlavi sources but subsequently concentrated almost exclusively on Hellenistic works in Greek and Syriac. (It is this program that is often described in popular histories as the "founding" by some ʿAbbāsid caliph, generally al-Ma'mūn or Hārūn al-Rāshīd, of an institution called the "House of Wisdom" (Arabic *bayt al-ḥikma*). In context, this term referred to an official court library or archive such as already existed under the Sasanians, rather than a particular new institute or academy dedicated to the scientific translation movement.)

The Arabic and Persian corpus of Hellenistic sources was further developed by Islamicate mathematicians, mostly Muslim but also including Jews, Christians, and Harranian/Sabian pagans. Nowadays, often subsumed under the term "Greco-Islamic science," this body of knowledge was a highly influential locus of medieval mathematical activity. Beginning in the eleventh and twelfth centuries, many Greco-Islamic works became known to the Latin world through translations from Arabic by scholars such as Adelard of Bath, Hermann of Carinthia, and especially Gerard of Cremona. Translations from Arabic also informed Byzantine science on subjects such as post-Ptolemaic astronomy and astrology or the decimal place-value "Indian numerals," although the Byzantines maintained their own Greek textual tradition of ancient Hellenistic works.

Farther afield, under the Yuan dynasty in the thirteenth century, the Muslim astronomer Zhamaluding or Jamāl al-Dīn (al-Bukhārī) joined the Chinese imperial Astronomical Bureau to carry out and supervise calendric compilations based on Greco-Islamic mathematical astronomy alongside indigenous Chinese astronomical systems. And in India, from about the fourteenth through eighteenth centuries under Tughluq and Mughal emperors, Arabic or Persian versions of a number of books by Hellenistic mathematicians and their Islamic successors were adapted into Sanskrit. In turn, imperial court scholars produced Persian translations of many Sanskrit classics, among them some canonical Indian mathematical works.

The sixth-century Roman scholar Boethius promoted the adoption of the four Greek sciences of arithmetic (number theory), geometry, astronomy, and harmonics as the Latin "quadrivium" (discussed more fully in Chapter 3), accompanying the "trivium" of grammar, logic, and rhetoric. Believing that this learning should be based upon a systematic Latin translation program of Greek philosophy and science including the entire Aristotelian corpus, Boethius himself produced translations of some mathematical texts. But his transmission project, lacking institutional or collegial support, did not survive his death. Large-scale Latin translation of Hellenistic mathematical sources in (a Byzantine form of) their original Greek was deferred until the European Renaissance.

THE "DARK" AND "GOLDEN" AGES OF MEDIEVAL MATHEMATICS

The same traditional historiography that pronounced the medieval period a Dark Age for the mathematical sciences of Europe often hailed the same period, or part of it, as a Golden Age for mathematics elsewhere. For Chinese mathematics, this designation is frequently assigned to approximately the thirteenth–fourteenth centuries, as well as to the earlier period in the third–fifth centuries sometimes called its "first golden age." The term is similarly applied to the sixth through twelfth (or even up to the sixteenth) centuries in the case of India, and to the eighth through twelfth or fourteenth centuries for Islamicate science.

The Renaissance and Enlightenment stereotypes of medieval European thought severed from its heritage of classical learning by barbarian violence and stagnating in an ignorant priest-ridden "dark age" are too familiar to need further exegesis here, and we leave it to the following chapters to rebut them. In this section we will touch on some of the remarkable achievements, and the national and cultural pride in their rediscovery, that have inspired the application of the "golden age" concept to medieval mathematical traditions beyond Europe.

From early medieval times in China, a comprehensive system of government bureaucracies oversaw the control of information generally, including the creation of calendars and almanacs, the production of dynastic histories, and the dissemination of Confucian classics and religious works such as Buddhist and Taoist writings. The national civil-service examination system for appointees to these bureaucracies required candidates to memorize a corpus of classical texts and demonstrate ability to reproduce their content, as well as mastery in composing verse and prose relating to various types of administrative documents. In mathematics, the chief of these canonical texts was the ancient classic *Nine Chapters*, whose eighth chapter is devoted to "rectangular arrays" where positions in different regions on the counting board were used to represent unknown quantities. The algorithms developed for these "rectangular arrays," to solve sets of simultaneous linear equations in as many unknowns as a given problem might involve, led Chinese mathematicians to introduce the concept of negative numbers and special rules for their arithmetic properties. These methods would receive their greatest generalizations during the extraordinary florescence of Chinese mathematics over about a half-century of the Song dynasty, beginning towards the middle of the thirteenth century and ending early in the fourteenth.

One of the earliest works of this period is Qin Jiushao's 1247 *Computational Techniques in Nine Chapters*, in which a general algorithm is presented for solving the Chinese remainder theorem using simultaneous congruences. This work also provides a general method for solving polynomial equations that is essentially the same as Horner's method devised nearly 600 years later for approximating the roots of polynomials. Li Zhi's 1248 *Sea Mirror of Circle Measurements* uses the *tianyuan* ("heavenly element") method whereby unknowns were used to provide algebraic solutions for 170 geometric problems, all related to one common diagram. What Li Zhi provides are algebraic solutions for polynomials up to the fourteenth power, including negative powers of the unknowns. Zhu Shijie's *Jade Mirror of the Four Unknowns* of 1303 deals with the elimination of unknowns from systems of polynomial equations in as many as four unknowns.

The preceding section of this chapter briefly alluded to thirteenth-century China's increased contacts with Islamicate countries and their sciences under the Mongols, exemplified by the career of the Muslim astronomer Zhamaluding. However, there is no definite evidence for any influence on this remarkable epoch of Chinese mathematics, with its emphasis on algebra and algorithms, from the algebra of its western neighbors. Certainly the terminology of the *tianyuan* method and the procedures themselves, carried out on specialized surfaces where positions of counting rods determined the unknowns to be manipulated, have no clear counterpart in any Arabic or Persian (or Sanskrit) texts.

In Indian mathematics the concept of a "golden age" is commonly attached to the period beginning with the work of Āryabhaṭa around 500, which we glimpsed in the previous section's discussion of the double distance method. Āryabhaṭa is the first of the great Indian mathematician-astronomers whose major treatise survives in its entirety, and like other Sanskrit mathematical treatises (or for that matter, the *Elements* of Euclid), it is more concerned with presenting a comprehensive systematized body of knowledge than with claiming or assigning priority for a particular result. Thus it is often impossible to distinguish with certainty between what Āryabhaṭa (or any later Indian mathematician) personally discovered and what he preserved from earlier writers.

But Āryabhaṭa's work indisputably provides the first known systematic treatment of arithmetic using decimal place-value numerals with zero, including square and cube root algorithms, and the first surviving procedures for constructing a table of sines, along with results in summation of series, solution of indeterminate equations, and other areas. Among other interesting features too numerous to mention, Āryabhaṭa's mathematical astronomy includes one of the few known assertions anywhere in the medieval world of the daily rotation of the earth, a view that was somewhat awkwardly explained away by his more cosmologically orthodox followers in subsequent centuries.

Other landmark figures most often named with Āryabhaṭa include his commentator Bhāskara I in the early seventh century, who stated an ingenious algebraic formula to approximate sine values, and Bhāskara's contemporary Brahmagupta, who recorded the first extant exposition of Indian algebra rules involving positive and negative numbers and zero. Brahmagupta is also renowned for, e.g., his statements on the properties of cyclic quadrilaterals, the solution of second-order indeterminate equations, and various interpolation and iterative-approximation formulas.

Dozens if not hundreds of other medieval Indian authors on mathematical sciences also contributed significantly to the development of the field. Probably the most famous of them all is the second Bhāskara or Bhāskarācārya ("Bhāskara the Teacher") in the twelfth century, whose comprehensive arithmetic, algebra, and astronomy texts remained canonical works in Sanskrit science for hundreds of years. Innovative aspects of these works include some trigonometric identities and ingenious infinitesimal arguments for some geometric formulas and for computing instantaneous orbital velocity. Bhāskara's books were among the Sanskrit texts selected for translation into Persian at the imperial Mughal court in the decades around 1600.

If the putative "golden age" for Indian mathematics is extended another two or three centuries, it embraces the brilliant work of Mādhava of Kerala circa 1400. Mādhava is remembered especially for his use of trigonometry, series, and infinitesimal concepts to deduce several of the infinite series expressions for trigonometric quantities that were rediscovered centuries later in European calculus, and which allowed him to compute an unprecedentedly accurate value of π. Within the next hundred years or so, prominent later members of the "Mādhava school" or "Kerala school" such as Parameśvara, Nīlakaṇṭha, and Jyeṣṭhadeva elaborated on Mādhava's work with, e.g., ingeniously argued demonstrations of his series results and modified spherical astronomy theories, including a "quasi-heliocentric" orbital model somewhat similar to that later proposed by Tycho Brahe.

The "golden age" frequently attributed to Islamic mathematics roughly coincides with the 'Abbāsid caliphate (750–1256), commencing with the eighth-century jump-start from translated Hellenistic texts mentioned in the previous section. Its best-known figure is probably that of the ninth-century Baghdad scholar al-Khwārizmī, today nicknamed the "father of algebra," whose *Book on Calculation by Restoration and Confrontation* (*Kitāb fī ḥisāb al-jabr wa'l-muqābala*) was one of the most influential Arabic works translated into Latin. The translators' attempts to render the names of the author and the treatise gave us the modern standard terms "algorithm" and "algebra" (Figure 0.3).

Al-Khwārizmī's contemporary, Thābit ibn Qurra, besides being a prolific translator of Greek and Syriac texts in Baghdad, is considered to have been the first to study the number-theoretic properties of what are now called "Thabit numbers." Several of their successors made significant advances in trigonometry (blending Greek spherics with Indian sines), Euclidean geometry, Ptolemaic geocentric astronomy, number theory, and algebra—among them the general solution of cubic equations by the Persian polymath Omar Khayyam. In Cairo around the start of the second millennium, Ibn al-Haytham, famous in Latin translation as "Alhazen," was a seminal figure in the field of geometric optics, besides writing on geometry, number theory, and other subjects. We have already seen in the preceding section some discussion of his contemporary fellow polymath al-Bīrūnī from Central Asia, who was renowned especially for his work in trigonometry and astronomy, not to mention history and ethnography. Naṣīr al-Dīn al-Ṭūsī in the thirteenth century is perhaps most famous for his "Ṭūsī couple," a geometrical device for modifying circular orbital motion, which was subsequently adopted by several European astronomers, probably including Copernicus.

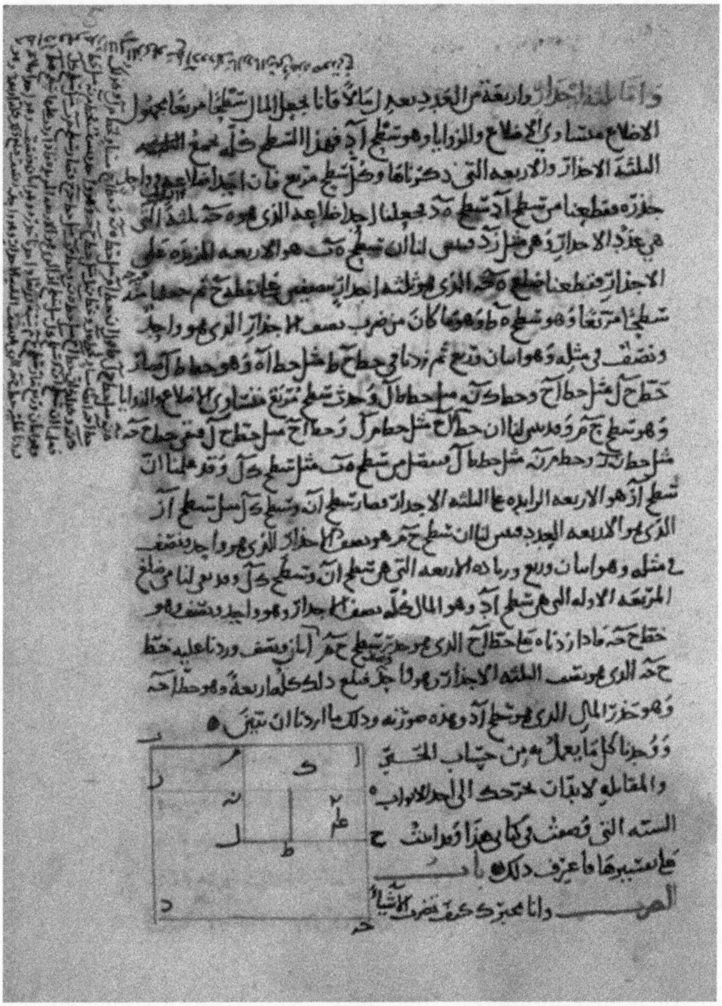

FIGURE 0.3. Geometric diagrams used to demonstrate solution procedures for quadratic equations, in a fourteenth-century manuscript of the ninth-century Arabic *Algebra* of al-Khwārizmī. Courtesy of the Bodleian Library, University of Oxford, Oxford, UK / Wikimedia Commons.

Of course, a "golden age" is seldom identified as such until long after it has ended. The question of what happened afterwards in the shift from the medieval to the early modern period will be explored in the final section of this chapter. Meanwhile, we briefly sketch the roles of mathematics in some of the fundamental ways that medieval people understood and interacted with the world they lived in.

THE SPHERES OF THE HEAVENS AND THE EARTH

The basic Hellenistic cosmological model of a spherical universe centered upon a likewise spherical earth, and the Greco-Babylonian coordinate systems and other mathematical tools used to quantify it, had an extraordinarily transformative impact on astronomy all the way from South and Central Asia to Western Europe. (In East Asian mathematical astronomy, on the other hand, traditional non-spherical cosmologies remained predominant until succeeded by modern heliocentrism.) Spherical orbital models and their accompanying mathematics enabled reasonably reliable prediction of complicated patterns of celestial phenomena even including solar eclipses, whose visibility is restricted to certain trigonometrically determined geographical locations.

Trigonometry, along with its concrete applications to observation by means of gnomon-shadows, astrolabes, and other instruments, was therefore the most crucial branch of mathematics for understanding the spherical cosmos. Apparently inspired by Hellenistic Greek trigonometry of chords, Indian astronomers sometime in the first few centuries of the Common Era developed the more efficient trigonometry of sines, eventually adding to it a plethora of formulas for computing various celestial and terrestrial coordinates.

The chord of an arc, and subsequently its half-chord or sine, was named in Sanskrit *jyā* or *jīvā*, words meaning "bowstring" (evidently because the chord and its subtended arc look like a strung bow). This latter term was rendered in Arabic as the consonants *j-y-b* and became conflated with the existing Arabic word *jayb*, meaning "fold" or "bay." European scholars' literal translation of Arabic *jayb* as Latin *sinus* is what gave us the modern standard term "sine" and its variants.

Islamic mathematicians combined Hellenistic spherics, Indian sine techniques, and new research to produce a major body of knowledge in plane and spherical trigonometry that was eventually assimilated into Latin mathematics and astronomy. The quantitative predictive power thus attained was unhesitatingly embraced by the mathematical practitioners responsible for maintaining calendar systems and identifying ominous signs in the heavens. Indeed, the predictive "power" most prized by the clients of medieval science was that of foretelling the future course of earthly events based on the configurations of celestial objects at crucial moments: i.e. the systems and methods of astrology and related forms of divination. In mathematical cultures assuming a spherical universe, the most complex and prestigious divinatory system was usually that of horoscopic astrology. Horoscopes could be cast not only for a nativity but for the time of some other momentous occasion, and interpreted in accordance with increasingly complicated models of interrelated planetary influences. Medical astrology or iatromathematics was a crucial

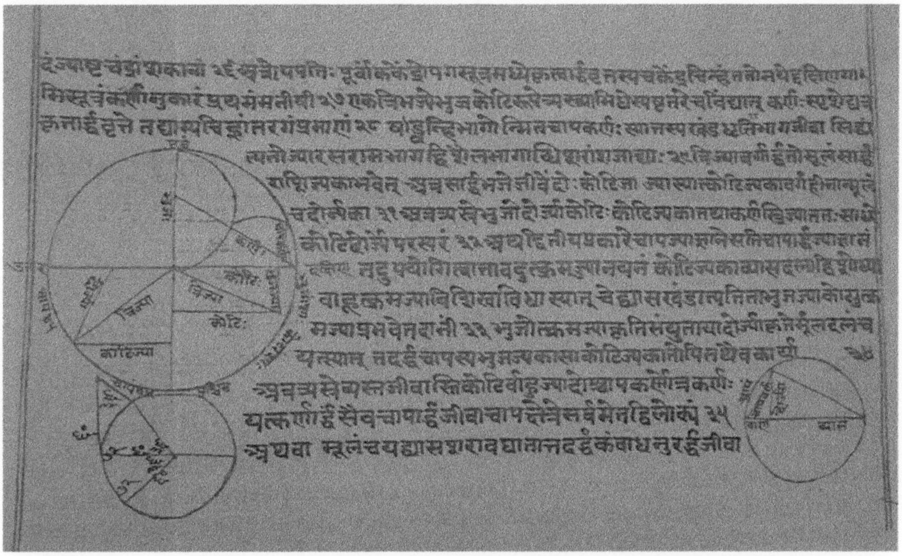

FIGURE 0.4. Expounding Indian trigonometry with a mix of standard Sanskrit technical terms and Greco-Islamic-style diagrams, in an 1846 manuscript of the 1639 Sanskrit *King of All (Astronomical) Treatises* by Nityānanda (RORI Alwar 2619, fol. 11v). Courtesy of the Rajasthan Oriental Research Institute, Jodhpur, India.

branch of this discipline, prognosticating the course of a disease or modifying treatment based on celestial omens.

Consequently, many medieval astronomer-mathematicians were also astrologers and/or physicians. An appointment as a court astronomer frequently involved both almanac compilation and astrological prediction, as illustrated by Islamic miniature paintings of court officials taking observations and casting horoscopes at the birth of a royal heir (as in Figure 2.2). And authors of mathematical astronomy works often wrote on astrology, too: they include, to sample just a few from the leading figures of medieval mathematical sciences already mentioned, Qutan Xida, al-Bīrūnī, Gerard of Cremona, al-Ṭūsī, and Parameśvara of Kerala.

The validity or legitimacy of astrological prediction was sometimes questioned in the medieval world, and various theological denunciations of astrological practices and practitioners were quoted from Buddhist, Hindu, Muslim, Jewish, and Christian texts. A minority of medieval mathematical practitioners expressed skepticism about some of astrology's underlying premises or specific methods, including some Muslim scientists such as Ibn Sīnā (Avicenna) and some Christian ones such as Nicole Oresme. Nonetheless, innumerable medieval Buddhists, Hindus, Muslims, Jews, and Christians—along with their contemporaries in many other faiths—sought out and accepted astrological guidance, further consolidating the identification of mathematical and astronomical knowledge with astrological practice.

In China, too, as Jean-Claude Martzloff points out, astronomers were responsible for "unceasingly reworked procedures devoted to the prediction of celestial phenomena and the calendar." The resulting innovations, including the introduction of non-decimal number systems and a unique notation for a written zero, differed significantly from the mainstream mathematical tradition mapped out in ancient classics like the *Nine Chapters*. Here, too, divinatory practices were closely associated with astronomy and the calendar: Martzloff notes a "strong link between predictive mathematics and divination and the predominance of empirical observation over theory" (Martzloff 2016: xxvii).

The culmination of the Chinese calendrical tradition in the medieval period was reached with the *Shoushi li* (Season-Granting Calendar), the official calendar system in China from 1281 until 1384, and one that was to have considerable influence on calendars adopted in Korea and Japan, too. When the Mongols came to power in China in 1271, they brought with them Greco-Islamic astronomy as well as the Muslim calendar created under the direction of the previously mentioned Zhamaluding to serve the large Arabic-speaking population of the empire. When the Han Chinese again took control of the country in 1368, the first Ming emperor Zhu Yuanzhang inherited both an Islamic and an indigenous Chinese Astronomical Bureau. In fact, at the collapse of the Yuan dynasty when the Mongol capital fell into Han Chinese hands, numerous confiscated Yuan astronomical works were deposited in the imperial library in the Ming capital of Nanjing, to be translated into Chinese at the command of the emperor. Thus spherical cosmology and its trigonometric methods became naturalized in Chinese astronomy as an adjunct to the orthodox Chinese systems.

Throughout this period, although many laypeople would naturally imagine the physical universe as described in sacred or other traditional cosmologies rather than abstract quantitative models, they nevertheless constantly participated in social structures based upon such models. The cycles of their daily lives were punctuated by official calendric and timekeeping systems, such as those prescribing the astronomical times and direction of Muslim daily prayers. Their expectations of the future were shaped by astrological prognostications, and their production and consumption of goods were partly determined by practices of geography and navigation on which transport of goods depended.

And of course, their buildings, their cities, their "infrastructure," and their wars were profoundly influenced by the application of mathematical expertise to physical objects in the terrestrial realm. Chinese engineering mathematics for the construction of city walls, dikes, ditches, moats, canals, watchtowers, pavilions both square and circular, and pyramids, to name only a few, goes back as far as the chapter on "Estimating workloads" in the ancient classic *Nine Chapters*. The role of mathematics in town planning is best reflected in the two major medieval capital cities, Luoyang and Chang'an (Xi'an), both of which were laid out with precise rectangular grids aligned on a north-south axis. An entire chapter of the oldest work on the architecture of government buildings in China, the *Standards and Models for Architecture* written in 1100 and printed in 1103, is devoted to general principles to be followed in constructing official buildings, including how to determine the correct directions for orienting them, checking horizontal and perpendicular alignments, calculating areas and volumes as well as the thicknesses and lengths of construction parts, etc. Later chapters of this work, whose author Li Jie was a Northern Song dynasty Director of Palace Buildings, explain how to calculate the requirements and overall cost of a given project in terms of time, labor, and the price of materials required for a given construction project.

A somewhat similar form of ancient "quantity surveying" is more lightly attested in the sections on practical computations in medieval Indian mathematical textbooks, where formulas are given for determining dimensions and amounts of, e.g., sawn lumber, heaped grains, stacked bricks, and excavations. The literature of *śilpa* (literally "skill" or "craft") contains detailed metrological specifications for the layout of various types of buildings including temples, the proportions of divine images to be installed in them, and so forth. However, the bulk of mathematical knowledge of this sort is probably to be found in the very understudied vernacular traditions of artisanal hereditary occupations in building and related areas, rather than in the academic Sanskrit treatises from which most of our understanding of Indian mathematics is drawn.

Much more is currently known about techniques of practical mechanics, engineering, surveying, etc. in the medieval Islamic world, largely through Arabic textual sources such as the treatises of the Banū Mūsā, Baghdad scientists of the ninth century, and al-Jazarī, an early thirteenth-century artisan-engineer. In these works there are descriptions of many machines including water-wheels, mills, trebuchets, water clocks, self-feeding oil lamps, and numerous automata and other "curiosity" devices. (Visitors to the Byzantine court in the tenth century reported the display of similarly elaborate automata there.) Historical and administrative writings describe some irrigation systems, water-tunnels, dams, and other public works, while surveying procedures are discussed in various texts ranging from al-Bīrūnī's astrolabe treatises to technical manuals on hydraulic engineering.

Surveying is one of the few practical-mathematics subjects for which we have significant knowledge of a body of texts and tools used in the early medieval

FIGURE 0.5. Back of a late fifteenth-century Persian astrolabe. A trigonometric quadrant for measuring angles is engraved in the upper left, while the bottom half contains two shadow squares for measuring distances and heights proportional to standard gnomon heights of 7 feet (*left*) and 12 digits (*right*). The lower rim includes two co-tangent scales, also in feet and digits; all the numbers are in alphanumeric (*abjad*) notation. Courtesy of the History of Science Museum, University of Oxford / Wikimedia Commons.

Latin West. These texts and tools were descended from the "gromatics" or "*groma*-use" knowledge of the "land-measurers" (*mensores*, *agrimensores*) of the ancient Roman Empire. (The *groma* was a surveying instrument used to set out straight lines and right angles.) By the eleventh and twelfth centuries, Latin "practical geometry" manuals began to circulate, such as the Parisian works mentioned in our previous discussion of the double difference method. This period also launched advances in Latin practical mechanics that resulted in the European mechanical clocks of the thirteenth and fourteenth centuries, soon followed by the innumerable ingenious devices of the Renaissance.

As the Latin medieval world drew to its close, it engendered another, more academic approach to the application of mathematics to physical reality. This was the study of matter and motion undertaken by the so-called "Oxford Calculators," a group of philosophers at Merton College in Oxford University in the fourteenth century. Thomas Bradwardine, Richard Swineshead, and their colleagues developed ways of analyzing uniformly accelerated motion, including relations between forces, resistance, and speeds. Much of their thinking invoked a fundamental notion of quantifiable variability known as the "latitude of forms." A major achievement of this school was the "mean speed theorem" articulated by William Heytesbury: namely, for a uniformly accelerating body, the distance it traverses while its velocity increases from an initial to a final value in a given interval of time is equal to the distance it would traverse in the same time if moving at a constant velocity halfway between the two values. This was demonstrated later in the century by Nicole Oresme at the University of Paris, who created a diagram of the motion treating the "latitude" of velocity and "longitude" of time as perpendicular coordinates. In this endeavor to rigorously mathematize notions of physical change, and the graphical tools of generalized latitude and longitude that accompanied it, we can discern—with the benefit of considerable hindsight—germs of the early modern "new sciences" of coordinate geometry, analysis, and mechanics.

TRANSITIONS TO THE POST-MEDIEVAL WORLD

In the mid-fifteenth century the human population, after having hovered between about 350 and 380 million for nearly three hundred years (including the plague-induced overall decline of the fourteenth century), began to rise again, surpassing 600 million by the early 1700s. The remarkable, and much remarked, story of mathematics in European learning during this period—its universities and academies, its Latin and vernacular printed texts, its blending of rediscovered ancient Greek works with its heritage of medieval Arabic sources to create the early modern subjects of algebra, coordinate geometry, heliocentric astronomy, classical mechanics, and calculus—is dealt with in Volume 3 of this series. In the remainder of this Introduction we explore what was happening in mathematical sciences outside the Latin world in the years between approximately 1400 and 1700, mostly in the aftermaths of what are often conventionally taken to be their various "golden ages." (However, since the total corpus of texts produced by these later traditions is at once very large, very diverse, and very understudied, the conclusions we draw about them must necessarily be quite tentative.)

FIGURE 0.6. Craftsmen setting boundary stones, in an early fifteenth-century manuscript of contemporary treatises on surveying and boundary marking by Provençal surveyor Bertrand Boysset. Photo by DeAgostini / Getty Images.

Renewed activity in the exact sciences in thirteenth- and fourteenth-century Byzantium included works on the number theory ("arithmetic") of the quadrivium; manuals for "Indian calculation" with decimal place-value numerals and for sexagesimal arithmetic; and numerous tables and treatises of mathematical astronomy and astrology, derived from both the original

Hellenistic Greek Ptolemaic corpus and its later Arabic, Persian, and Latin descendants. After the 1453 Ottoman conquest of Constantinople, Byzantine science remained active both in the Latin West (to whose classical renaissance it famously contributed many Greek manuscripts of Hellenistic works, as well as scholars who could read them) and in the Ottoman Empire. Ottoman Greece produced several scholars known to have written on mathematical subjects, including Damaskenos the Stoudite in the sixteenth century and Theophilos Korydalleas in the early seventeenth.

Islamicate mathematical sciences in the Ottoman Empire and its neighbors disseminated and built upon knowledge from the Ilkhanid, Mamluk, Timurid, and earlier dynasties. After the twelfth century, new institutions including several observatories and many endowed teaching centers or madrasas supplemented existing courts and mosque colleges as loci of scientific activity. The astronomer-mathematicians Naṣīr al-Dīn al-Ṭūsī in Maragha in the late thirteenth century, Ulugh Beg in Samarkand in the early fifteenth, and (for a brief time, as described at the start of this Introduction) Taqī al-Dīn in Istanbul in the late sixteenth century, all became renowned as founders and/or leaders of observatories generating substantial scholarship in mathematical astronomy and trigonometry. The emergence of the "science of timekeeping" (*'ilm al-mīqāt*) under the Mamluks starting in the late thirteenth century and the proliferation of official timekeeper positions tasked with computing the times of Muslim prayer at a particular locality, also expanded institutional support for aspects of astronomy and related mathematical subjects, including the design of observational and computational instruments.

Persian and subsequently Turkish continued to develop as important languages of mathematical sciences, along with the ubiquitous Arabic. All these languages left their mark also on the rich Ottoman textual record concerning the making of maps and architectural ornaments, whose manuscript remains (along with the magnificent tiling patterns of many mosques) preserve some of the most visually stunning representations of medieval Islamicate practical geometry. (See, for example, the intricate Islamic *muqarnas* depicted in Figure 6.4.) Persian and Arabic dominated the extensive mathematical literature produced during this time in Islamic polities to the east, including Safavid Iran and Mughal India.

In general, mathematical sciences in the Islamic world became further institutionalized within endowed schools (which provided stipends for students as well as teachers), timekeeping offices, and other official bodies. Serving the mathematical needs of increasingly extensive and populous Muslim-led states required a constant supply of competent practitioners. Generating this supply involved not only increasing professional entrenchment but a professional emphasis on pedagogy and disciplinary consolidation, experimenting with various combinations of canonical treatises by Greek (in translation), Arabic, and Persian authors with newly composed textbooks to shape instructional curricula.

Court scholars labored at the construction of improved astronomical handbooks (*zījes*) and comprehensive technical encyclopedias in honor of rulers, as we know partly from the many lavishly illustrated and decorated manuscripts of such works produced by palace libraries and prized as luxury commodities or diplomatic gifts. Although there were still numerous scholars who engaged in advanced mathematical studies with individual teachers and explored new paths of research, especially in planetary theory and instrument design, it is probably fair to say—at least until and unless ongoing historical research upends our current assumptions—that mathematical practice was largely dominated by the administrative and career requirements of their increasingly professionalized field.

Meanwhile, the institutional fortunes of the Sanskrit-based mathematical sciences practiced among South Asia's non-Muslim majority were more fluctuating. Many courts of the Tughluq, Mughal, and other Indo-Islamic dynasties employed Hindu or Jain officials as astronomer-astrologers or administrators, working with Muslim colleagues to coordinate quantitative practices involving finance, official calendars, prognostications, and so forth. The hundreds of Hindu rulers who governed smaller regions, either independently or under (actual or nominal) Muslim liege lords, of course also had court functionaries such as astrologers who were learned in Indian mathematical sciences. At both Muslim and Hindu courts some renowned works in Sanskrit mathematical astronomy were composed.

Muslims sometimes studied Indian *śāstras*, possibly including mathematical works, under Hindu teachers. And when the Mughal emperor Akbar in 1582 decreed that Persian would be the official administrative language of the empire, displacing the Hindi/Urdu dialects previously used by most non-Muslim administrative employees, he opened the Persian-medium madrasas to non-Muslim students. Many of these future clerks and accountants thus absorbed some of the Islamicate curriculum in arithmetic, geometry, and astronomy, among other subjects. On the other hand, when political conflict or palace controversy not seldom resulted in the destruction or plundering of a Hindu or Jain temple (either by Muslims or by Christians during the Goa Inquisition), the indigenous educational and scholarly activities associated with it generally suffered.

Throughout, India's innumerable familial and occupational lineages in various forms of mathematical practice continued to teach new generations and produce new works and commentaries. A few examples of this heritage include the Jain astronomer Mahendra Sūri, who in 1370 composed for the Tughluq emperor the first Sanskrit treatise on the Islamic astrolabe; Jñānarāja of Maharashtra, author of a 1503 mathematical astronomy treatise modeled on that of the great Bhāskara II; Gaṇeśa from Nandgaon near Mumbai, whose 1520 astronomical handbook became one of the most popular Sanskrit works in its genre; Nityānanda in the early seventeenth century, who produced at the request of Shāh Jahān's minister a Sanskrit translation of a Persian *zīj* as well as another substantial treatise combining Indian and Islamic models; and various members of the astronomical team at the Jaipur observatory of Jai Singh II in the early eighteenth century, who produced with Muslim and European colleagues Sanskrit translations of Persian and Latin works, besides creating other treatises in Sanskrit and designs for monumental observing instruments.

Also worthy of note are several distinguished mathematical families working in Varanasi (Banaras) in the sixteenth and seventeenth centuries, some descended from scholars who had relocated from western India (among them a former pupil of the abovementioned Gaṇeśa) and other areas to this ancient center of Hindu learning. And last but far from least, there was the continued work of the successors of the fourteenth-century Mādhava of Kerala mentioned in a previous section, who continued an unbroken chain of mathematical writing and teaching until at least the early seventeenth century. The larger intellectual context of all the foregoing accomplishments involved many aspects of re-engagement with classic treatises in the field: for example, extensive new commentaries exploring proofs and philosophical inferences, the proliferation of new forms and formats of table texts to streamline astronomical computations, and linguistic diversification into non-Sanskrit vernaculars.

In China, the Ming dynasty returned rule of the country from the Yuan to the Han Chinese in 1368, and by 1600 its population had increased to nearly 200 million. But economic and agricultural failures, as well as the 1633–44 Great Plague that devastated much of China, eventually led to the downfall of the Ming and its replacement in 1644 by the Manchu Qing rulers, the last of China's imperial dynasties.

By the Ming period, counting rods as the basic tools of computation had been replaced by the bead abacus or *suanpan*, a device better suited to mercantile reckoning than to the abstract methods of the abovementioned *tianyuan* polynomial algebra. In fact, the *tianyuan* methods rapidly ceased to be developed, or even understood, in fourteenth-century Chinese mathematics. Although some works on traditional mathematics (especially expositions of the ancient *Nine Chapters*) continued to be produced along with descriptions of *suanpan* calculation, they broke no new ground in the subject. Nor did the great Ming encyclopedia, the *Yongle dadian* completed in 1408, whose 22,937 rolls were intended as a comprehensive collection of Confucian, Buddhist, Taoist, and Divination works, as well as all that had been written in history, philosophy, arts, and sciences. Produced (and incompletely surviving) only in a few handwritten manuscript copies, this monument to the preservation of past knowledge contains two volumes of fragments of the *Nine Chapters* and other mathematical classics, but apparently failed to inspire further advances.

The Yuan legacy in mathematical astronomy, especially the activities of its Islamic Astronomical Bureau working in parallel with the scholars of indigenous

Chinese astronomy, met with greater receptiveness in the Ming era. The first Ming emperor merged the Islamic and Chinese institutions, recruited additional Muslim astronomers, and set his staff to translating some of the Islamic treatises into Chinese. The division of computational labor between the two very different systems was a complicated process, and nothing like a complete synthesis of them had been achieved by the late sixteenth century when Jesuit missionaries in China began to introduce contemporary European versions of astronomy.

These exchanges between Chinese and Jesuit scholars also re-energized Chinese engagement with non-astronomical mathematics. By the end of the seventeenth century, these international collaborations had produced Chinese translations of some of the books of Euclid and numerous modern Latin works on practical arithmetic, surveying, geometry, logarithms, trigonometry, spherical astronomy, and the Western astrolabe. In the eighteenth century, Chinese intellectuals launched a concerted effort to collect and study the ancient classics of traditional Chinese mathematics, partly to prove that indigenous forms of the learning in the "Western" mathematical texts introduced to China by early modern Jesuits had been known much earlier, although subsequently forgotten. Polymaths like Dai Zhen and Kong Jihan were leading figures in this effort to reclaim the great discoveries of China's mathematical past.

When we zoom out to try to attain the broadest view of mathematical sciences east and south of the Mediterranean in the mid-second millennium, the general outline we dimly perceive is that of textually mature disciplines responding to the expansion of administrative functions and professional opportunities in growing societies and states. Accountants, astronomers, astrologers, surveyors, timekeepers, teachers, and many more had crucial roles to play in this growth, with established bodies of specialist knowledge to guide their thinking. They also had scope for exercising technical virtuosity in innumerable refinements of theory, technique, and instrumentation based upon this knowledge.

Some of the greatest demands upon mathematical practitioners' virtuosity involved the synthesis of markedly different theoretical approaches. Such integrative projects might be comparatively modest in scope, like the work of the Iranian astronomer Shams al-Dīn al-Khafrī in the early sixteenth century to reconcile and refine his predecessors' orbital models for the moon in accordance with the physical principle of uniform circular motion. Or they might entail ambitious efforts at large-scale cross-linguistic amalgamation, such as the Ming Astronomical Bureau's attempted unification of traditional Chinese and Islamic calendar systems in the fourteenth and fifteenth centuries; or the "sinification" of European mathematics and astronomy by Jesuit and Chinese scholars in the seventeenth; or Maharajah Jai Singh's undertaking to combine and compare indigenous Indian, Islamic, and European astronomical systems, based on data collected in his newly constructed observatories, in the early eighteenth. (Technical translation and the creation and reconciliation of different languages' scientific

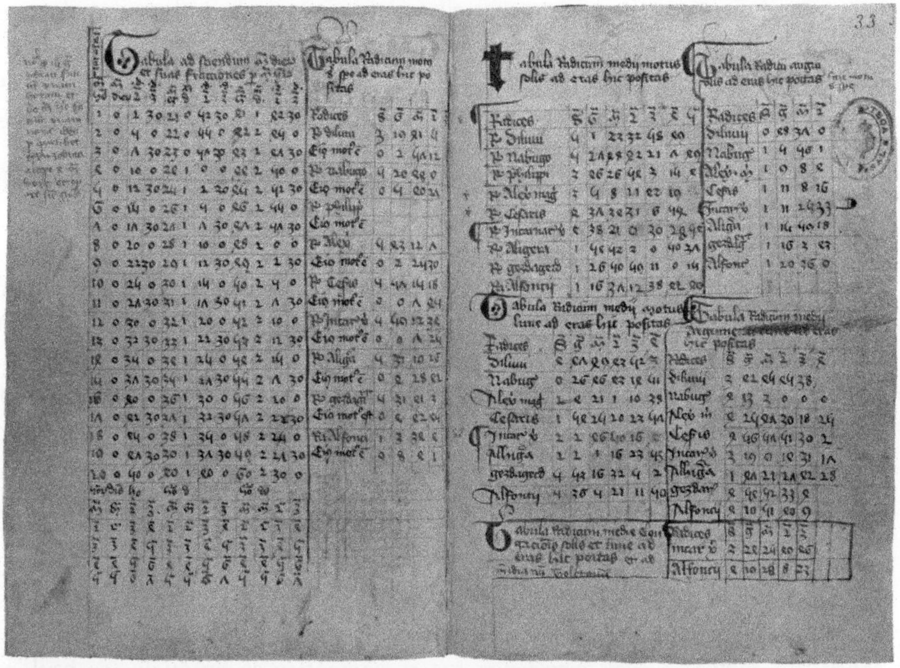

FIGURE 0.7. Latin astronomical tables in the Parisian-Alfonsine tradition with decimal place-value numerals, in a manuscript from early fifteenth-century Spain. The leftmost table converts minutes (which were written in red) into sexagesimal fractions of a day (in black); the other tables contain planetary mean motion parameters for various epochs (MS Madrid, Biblioteca Nacional de España 4238, fols 32v–33r). Courtesy of the National Library of Spain, Madrid / Alamy Stock Photo.

vocabularies were no inconsiderable tasks in their own right. They challenged the ingenuity of mathematical authors not only in the above-mentioned instances involving, e.g., Turkish and Indic vernaculars, but in many other contemporary cross-linguistic transmissions that we have no space to discuss, such as those from Chinese to Japanese or from Arabic to Malay.) All these attempts at mathematical synthesis and consolidation required practitioners to make sense of unfamiliar material, some of which apparently contradicted their existing knowledge. Often there was no easy way to determine which of two different approaches would be preferable, especially in the case of orbital models in astronomy whose empirical assessment might require extensive programs of observation.

Over time, more and more of these interactions involved some aspect of the Latin mathematical sciences of Western Europe, generally in a heliocentric theoretical framework. Such encounters were not typically perceived on the non-Latin side as "revolutionary" clashes of fundamentally incompatible philosophies, much less of religious faiths (although the latter interpretation was sometimes encouraged by evangelizing missionaries). Rather, the ideas of "Firangi" or European authors such as Copernicus, Clavius, Napier, and Duret, like the ideas of so many other learned scholars of other times and places, were regarded as a potential source of useful material to be incorporated into existing mature scientific knowledge systems.

The question is often asked why Western Europe's post-medieval mathematical renaissance or revolution was not independently paralleled or anticipated in contemporary East or South Asia or elsewhere in the Islamic world. As noted above, historical knowledge of these cultures' mathematical activity in this period is still extremely incomplete. And the knowledge of it that we do have suggests a densely intertwined growth of technical and social developments whose surface the brief descriptions in this section have barely scratched.

Nonetheless, many histories of science have attempted to tackle this question by hypothesizing various causal or contributing factors, whether technical, cultural, social, or economic. It has been speculated that cultures beyond the Latin West were prevented from attaining the mathematical breakthroughs of the Scientific Revolution because, for example, their societies were under the sway of religious orthodoxy or collectivist attitudes; or because they did not share the ancient Greek reverence for mathematics as uniquely truthful or cosmologically fundamental; or because they did not have the Protestant work ethic, or they did not have movable type, or they did not have gunboats, or they did not have capitalism. Perhaps they were too much engaged with occult sciences like astrology and magic (whose imaginary mathematical models for mysterious unseen forces fostered irrational superstition), or they were not engaged enough with occult sciences like astrology and magic (whose imaginary mathematical models for mysterious unseen forces prefigured the invention of mechanics). And so on, over a vast range of possible influences potentially shaping the course of mathematical thought. Any one of these proposed reasons will be applicable to some of the historical realities and inapplicable to others; taken together in any combination, they explain anything and nothing.

Returning our gaze to the enigmatic figure of Taqī al-Dīn in 1580s Istanbul as he contemplated the ruins of his hard-won observatory, we might be tempted to infer that its fate symbolizes a similar sort of inherent cultural obstacle—say, religious hostility to science, political despotism, palace machinations—that inevitably blocked Islamicate astronomy's path to becoming a modern science on a par with that of the Latin world. Yet, if demolition of observatories were a reliable indicator of scientific collapse, we would expect Christian IV's 1601 destruction of Tycho Brahe's Uraniborg and Stjerneborg to have likewise derailed Europe's Scientific Revolution. We must look a little more carefully at the account in Lokman's Persian poem:

> . . . the Potentate who is the defender of the Religion spoke thus to his astronomer, Taqī al-Dīn:
> "People of learning have made inquiries concerning this: Oh you witty man of conscientiousness and perfection,
> Inform me once more on the progress and the results of observations . . ."
> Taqī al-Dīn answered, "In the *zīj* of Ulugh Beg
> There were many doubtful points, oh exalted King;
> Now through observations the tables have been corrected . . .
> From now on, order the abolishment of the Observatory,
> To the consternation of the ill-wishers and the jealous."
> . . . They pulled the Observatory up by the roots
> And made subside all the work concerning the firmament . . .
> Do not make decisions concerning the affairs of the firmament.
> For who, besides God, knows the gait and the revolution of the heavens?
> The influences of the higher spheres and their appearances are, without any doubt, innumerable.
> Plato who was the master of these arts was bewildered in the face of such thoughts.
> Aristotle, Hippocrates, and Socrates too . . .
> They were unable, however, to penetrate that mystery
> Which extends from the inner depths of the earth to the ninth sphere.
> What hope then can you have of uncovering all these matters,
> That you make diversions from the surface of the earth and indulge in celestial affairs? (Sayili 1956: 481–4)

Several of the above lines (although, curiously, never the ones expressing Taqī al-Dīn's own acquiescence in the observatory's fate) have sometimes been quoted in popular histories of science to deplore the apparent defeat of empirical rationality by an orthodox dogmatism which, it is suggested, condemned Islamic science to centuries of decline or even eclipse. However, we might also read them as attesting to a growing theological-philosophical distrust of the predictive claims of astrology, with its attempts to deduce events on the "surface of the earth" from the "mystery" of the "influences of the higher spheres." We might also recall that the stated purpose of a program of astronomical observations, according to the poet, was to revise obsolete works and remedy errors and doubts in order to "become freed from observation programs until doomsday." Anathema though it may seem to modern scientists, an observatory of the period may have been viewed by its sponsors more like a present-day grant, a temporary public work that in practical terms could be considered expendable once its research intentions had been carried out.

Other sources and events that we encounter in the history of pre-modern science similarly remind us that there is seldom a single cause or a simple explanation for any particular development. The chapters that follow aim to do justice to as many of the manifold complexities of medieval mathematics as we can manage here, realizing that inextricably woven together are its many triumphs and declines, its darkness and its gold.

Needless to say, the task of describing with reasonable accuracy in seven chapters fundamental aspects of mathematics in human society over more than a thousand years of development in no less than six major classical languages has been a daunting one. The three editors of this volume would like to express our deep gratitude and admiration to our three colleagues who accepted our invitation to share the challenges and labors of writing it. The invaluable contributions of Jian-Ping Jeff Chen in Chapter 5 and Sonja Brentjes and Nathan Sidoli in Chapter 7 have formed part of our collaborative process of revision and editing in the cause of continuity and consolidation. Thus it would be fair to say that no one contributor is solely responsible for any given chapter, each of which has been cooperatively shaped by points discussed, rewritten, or imported from elsewhere in the volume.

We have also benefited from the valued scholarship and advice of innumerable fellow researchers in the mathematical sciences of the medieval period, whose guidance we cannot adequately acknowledge within the limited space allotted to us for notes and bibliographies, but to whom we are likewise lastingly grateful. Any inaccuracies and omissions that remain despite their help and our best efforts are due either to the intrinsic impossibility of trying to cover as much of the monumental richness of medieval mathematics in one volume as we could, or to our own errors of judgment, or both.

We also want to thank the many institutions, archives, libraries, and other repositories that have allowed us to use images from their collections, in the majority of cases *gratis*. And lastly, we are indebted to our editors at Bloomsbury, Kakul Ehsan Butt, Roza I. M. El-Eini, Elizabeth Holmes, Merv Honeywood, and Tristan Palmer, for their careful work, especially in the final stages of this project. To everyone who has contributed to this enterprise in so many different ways, we are sincerely grateful.

CHAPTER ONE

Everyday Numeracy

A Thousand Years of Reckoning

JOSEPH W. DAUBEN, CLEMENCY MONTELLE, AND KIM PLOFKER

The trade networks of the Old World that flourished in late antiquity began to break up as the middle of the first millennium approached, with dynastic fracture in China and the weakening of Roman imperial power. State sponsorship of learning contracted in some regions—including Rome where most of the public libraries of the early empire were closed by the fifth century—and expanded elsewhere, as in the Sasanian and Gupta empires dominating much of West and South Asia. Medieval cultures still retained many devices and practices of popular numeracy in the ancient world, such as Chinese or Roman counting boards and Greco-Roman dactylonomy (finger-reckoning). These were codified in pedagogical and administrative structures that developed in diverse new directions even as they consciously looked back to their ancient heritages.

No encyclopedia, much less a single chapter, could do full justice to the variety and complexity of this evolution of numeracy in hundreds of interlinked societies and at least a dozen scholarly languages over the course of a long millennium. Here we attempt only to sketch some salient and widely shared features of numerate practices in medieval everyday life, as far as the limited extant sources (and our own limited knowledge of them) permit. Beginning with the most basic aspects of near-universal forms of quantification in language, we turn to its applications and techniques within everyday business of household and trade. Then we examine some of the educational structures and "mathematical folklore" that helped maintain numeracy among people outside professional mathematics. Finally, we trace the cross-cultural history of the medieval development that had possibly the most revolutionary impact on popular numeracy in the post-medieval world west of China: namely, the emergence and transmission of the now-universal decimal place-value number system.

COUNTING AND QUANTIFICATION

The ubiquitous everyday practices of counting and measurement in medieval communities supported a rich variety of metrological norms and number representations. Instead of a unified set of standard techniques, different systems flourished side by side, both within and beyond shared cultural and linguistic traditions. Merchants, for example, had to understand not only the relationships between units of money or weights and measures in different systems, but also the rudiments of various calendar reckonings observed by their users. In many scripts, alphabetic numerals and specialized fraction glyphs, not to mention standard patterns of tally marks, coexisted with other ways of writing numbers.

Counting and computation in daily life were heavily marked by orality. Traces of medieval enumeration practices may persist in many surviving counting-out rhymes and other doggerel, such as the twenty Celtic-derived number words in the traditional British shepherd's counting verse beginning *yan tan tethera* . . . ("one, two, three . . ."). Formal systems of counting were usually set out in the introductory sections of mathematical texts and primers, but were also the focus of attention in other literary genres, such as encyclopedias and grammar texts. The sixth-century scholar Isidore of Seville included in his knowledge compendium *Etymologies* a section dedicated to mathematics which opens with the origins of number words. His justifications for number names are mostly based on his own speculations about their etymological origins from Greek or Latin words with similar meaning or pronunciation:

> The term "coin" (*nummus*) gave its name to "number" (*numerus*) and from its pervasiveness, it bestowed the name. "One" (*unus*) draws its name from Greek; for the Greeks call one ἕνα. Thus it is for "two" (*duo*) and "three" (*tres*), which the Greeks call δύο and τρία. "Four" (*quattuor*) takes up its name from the square (*quadratus*), but "five" (*quinque*) did not take its name from nature, but rather from the arbitrary will of the one who bestowed the names upon numbers. "Six" (*sex*) and "seven" (*septem*) come from the Greek. For we pronounce an *s* instead of the rough breathing sound (i.e. the *h*-sound) in many words that are aspirated in Greek. Thus, we have *sex* for ἕξ and *septem* for ἑπτά just as we say *serpillum* (i.e. a type of thyme) for the herb *herpillum*. But "eight" (*octo*) is a direct translation, as we and the

> Greeks say it the same way. Similarly, they say ἐννέα, we say *novem* (i.e. "nine"); they say δέκα and we say *decem* (i.e. "ten") . . . (Barney et al. 2006: 89–90)

Isidore's list of number names concludes with the "myriad," which he incorrectly identifies as the Greek work for "thousand." The ancient *myriad* was in fact 10,000, the largest number specifically named in common usage in Greek antiquity.

Similarly, in ancient and medieval China, quantities such as 10,000 or 100,000 served as a practical upper limit on enumeration. However, the system of Chinese numbers was such that it was easy to accommodate numbers that were arbitrarily large. For example, *The Mathematical Classic of Sun Zi* (no later than the fifth century CE) lists the following means of naming large numbers in an ascending scale of powers of ten. The scale begins with one 萬 *wan*, 10,000, so that *wan wan* is 10,000 10,000 or 10^8. Larger numbers followed in an orderly ascending sequence:

> 10^8 萬萬 *wan wan* is called 億 *yi*
> 10^{16} 萬萬億 *wan wan yi* is called 兆 *zhao*
> 10^{24} 萬萬兆 *wan wan zhao* is called 京 *jing* . . . (Lam and Ang 2004: 193)

This eventually led to a number Sun Zi listed as 萬萬正 *wan wan zheng*, called 載 *zai*, 10^{80}.

Another medieval Chinese mathematical text, the (probably sixth century CE) *Memoir on Some Traditions of the Mathematical Art*, introduced three levels of large numbers, ranging from a "lower" and "middle" to "upper" level, the largest number of which is 10^{4096}. This text also betrays a close affinity to Buddhist texts and even incorporates language from a Sanskrit classic, the *Śārdūlakarṇāvadāna*. Its term for an infinitesimally small period of time, a split second or "twinkling of an eye," *kṣaṇa* in Sanskrit, appears in the *Memoir* as *chana*.

The Iranian polymath Abū Rayhān al-Bīrūnī commented on such linguistic upper bounds for number names (ignoring the Syriac and Arabic specialist term *ribwa*, an exact equivalent of the Greek myriad):

> In arithmetic all nations agree that all the orders of numbers (e.g., one, ten, hundred, thousand) . . . that each order is the tenth part of the following and the tenfold of the preceding. I have studied the names of the orders of the numbers in various languages with all kinds of people with whom I have been in contact, and have found that no nation goes beyond the thousand. The Arabs, too, stop with the thousand, which is certainly the most correct and the most natural thing to do . . . Those, however, who go beyond the thousand in their numeral system are the Hindus, at least in their arithmetical

> technical terms . . . They extend the names of the orders of numbers until the eighteenth order [10^{17}] . . . (Sachau 1910: 174)

The introduction of number words in European languages to name much larger quantities than the myriad seems likely to have been prompted by the increase in volume of trade commodities associated with the late medieval commercial revolution. The Italian term *millione* first appeared in print in the fourteenth century to represent "a great thousand" (i.e. a thousand thousand, the modern million), and subsequently spread across medieval Europe.

Everyday reckoning with small quantities, or fraction arithmetic, was likewise strongly shaped by ancient practical tradition. The Greco-Egyptian unit fractions and the duodecimal fractions of the Romans persisted in medieval Europe and Islam. A more general system of decimal "common fractions" with any integer numerator or denominator, written in the numerator-over-denominator format still employed today, was adopted from Indian arithmetic by authors in the Islamic world and subsequently in Europe. The simplest fractions such as a half or a third served for rough estimations as well as precise divisions, as illustrated by Chinese terms for one-third meaning literally "small half" and for two-thirds meaning "large half."

Speculations in religious and philosophical traditions about quantities that might be uncountably big or unfathomably small often involved mundane descriptions to motivate their scope and scale. For instance, a popular Buddhist text from the early Common Era, the *Lalitavistara sūtra*, contains a passage on an exchange of mathematical challenges between the Buddha and a mathematician called Arjuna. In one of these Arjuna asks the Buddha to compute the hundredth power of 10 million: in response, the Buddha precedes to name all the orders of magnitude from 10 million to its hundredth power. He goes on to link these names with concrete quantities such as the height of the cosmological mountain Meru or the grains of sand that could fill the bed of the river Ganges—or even of 10 billion such rivers. The Buddha likewise expounds the smallest conceivable quantity by means of comparisons linking everyday objects in order of increasing size:

> Seven of the smallest particles is one small particle. Seven small particles is one water particle. Seven water particles is one airborne dust particle. Seven airborne dust particles is one dust particle on a hare. Seven dust particles on a hare is one dust particle on a sheep. Seven dust particles on a sheep is one dust particle on a cow. Seven dust particles on a cow is one louse egg. Seven lice eggs is one mustard seed. Seven mustard seeds is one barley grain. Seven barley grains is one finger joint . . . (Dharmachakra 2013)

A somewhat similar interest in articulating the physical nature of the very small appears in theological speculations on indivisibly minute divisions of time by the eighth-century English monk Bede:

> to suggest the swiftness of the Resurrection, stating *We shall all rise, but we shall not all be changed, in a moment, in the twinkling of an eye* . . . many writers indiscriminately call that tiniest interval of time in which the lids of our eyes move . . . and which cannot be divided or distributed, either a *momentum*, a *punctus*, or an atom. (Wallis 1999: 16, original italics)

These and many more such examples illustrate medieval willingness to explore ideas of magnitude beyond ordinary finite quantity, and to connect them with real physical experience outside the abstractions of specialist mathematics.

Non-Decimal Number Systems

Despite al-Bīrūnī's confident assertion about the universality of the base-ten principle in number systems, non-decimal alternatives persisted in a variety of everyday contexts. Vigesimal or twenty-based counting systems can be found in the Celtic and Germanic as well as Romance branches of the Indo-European language family. Even some modern French number words, such as *quatre-vingts* ("four-twenties") for eighty and *quatre-vingts-dix* ("four-twenties and ten") for ninety, hearken back to Old French medieval vigesimalism, which also spawned many similar (now archaic) number words including *vint e dis* ("twenty and ten") for thirty and *deus vins* ("two twenties") for forty. In late Old English, number words for sixty and eighty were formed by adding the Norse suffix *-score* to "three" and "four." In Danish, beginning around the fourteenth century, a different "times twenty" suffix generated terms such as *fiyrsin tiughæ* ("four times twenty," later *firs-indstyve*) for eighty and *half fæmpt sin tiygh* ("half-fifth times twenty," later *halvfemsindstyve*) for ninety.

A glance at the larger medieval world reveals an even richer diversity of enumeration systems, ranging from the vigesimal structure of Mesoamerican bar-and-dot numerals and Inupiaq (Eskimo-Aleut) number words to the base-four number words of the Papuan Kewa language. Within the major literate traditions of Eurasia, the single most important non-decimal system was the base-60 or sexagesimal positional numbers (both integer and fractional) applied to time and spatial coordinates. Sixty-based cycles were used in Chinese and Indian calendrics long before their astronomers encountered the horoscopic astrology and spherical astronomy founded on late Mesopotamian-Hellenistic models. But it was ultimately from those models that various medieval metrological systems in Greek, Latin, Sanskrit, Arabic, Persian, and many related languages derived their suites of base-60 units for measurement of time and/or degrees (360ths of a circle). Such units were familiar to Chinese astronomers in the Tang and Ming dynasties who worked with Indian or Islamic astronomical sources, but they did not become common in everyday life in China until after the introduction of modern European astronomy and horology.

Units of Measurement

Medieval everyday numeracy depended not only upon abstract numbers but also on a profusion of concrete terms to quantify specific types of material items. Some were just lightly concretized number words, such as the various European cognates of "dozen" meaning "a group of twelve" (from Latin *duodecim*). Some, especially in East Asian languages, were numeral classifiers assigned to a particular category of counted entity; thus a number-word like "five" would be differently modified depending on whether it referred to five persons or five books, for example. Some were specialized terms indicating both cardinality and materiality, such as English "ream" (derived from Arabic *rizma*, "bundle") meaning a certain number of sheets of paper. The cardinality of a particular term might even change depending on what it measured. For example, a late medieval English treatise on weights and measures declares the "hundred" of, e.g., herrings or ells of linen cloth to be equal to 120 (the Germanic "long hundred"), while a "hundred" shillings means 100. (It further states that a "hundred" is 160 for counting dried fish but 108 for pounds of wax, sugar, pepper, etc.)

The most prolific mathematical terms in every medieval society were the endlessly diverse units of measure for continuous quantity. Such units might intrinsically vary from individual to individual, like the foot-length, finger-breadth, and forearm-length or cubit. Their standardization, if any, might be specific to a particular region or era or type of goods. For instance, the term "ounce" (from Latin *uncia*, one-twelfth of a Roman *libra* or pound) signified both the "troy ounce" used especially in weighing precious metals, costly medicines, etc. and the somewhat lighter ounce of the late medieval "avoirdupois" system used to weigh bulky goods such as wool. The length measure *dhirāʿ* in the Islamic world, originally based on the ancient forearm-cubit, morphed into at least several dozen different conventional lengths depending on whether, for instance, their users were carpenters or weavers, or weavers of Cairo as opposed to those of Damascus, etc. This bewildering range of variability was also seen in many other units of medieval metrology.

Rulers, guilds, and other institutions pushed back against this endless proliferation of units with various attempts to impose consistency. Legal decrees, public displays of standard measures, and official acts of enforcement all played a role in these efforts. Common standards were often at least nominally based on the ruler's personal measure, such as the "royal cubit" (*pārameśvarīya-hasta*) mentioned in the dimensions of a land grant described in a ninth-century temple inscription in Gwalior in central India. A physical model of an official standard was useful where high concentrations of trade enabled easy access to it: e.g., the iron rod whose length marked the official ell of the great cloth market at the summer fair at Troyes in late medieval France, against which all the vendors' measuring sticks had to be tested.

Enforcement of metrological compliance was not always so rigorous, although some regulatory practices aspired to it. Under Islamic law, for example, the official *muḥtasib* or inspector of markets (and general guardian of public morality) was charged among numerous other duties with detecting and punishing vendors who cheated customers in any way, including the use of false measure. There are records of *muḥtasib* officials having bakers and sweet-sellers in Fatimid Cairo beaten and publicly humiliated for giving short weight on their goods. Likewise, the tenth-century Byzantine Greek *Book of the Prefect* includes among its rules for the governance of silk-dealers by the Prefect of Constantinople the injunction: "Whoever possesses scales or weights not marked with the seal of the Prefect is to be beaten and shaved."

Quantitative units of time were less subject to (although not totally exempt from) such localized variation, given that a period such as a month or day or hour was ultimately regulated by celestial events immune to dishonest tampering. Its passage would typically be demarcated by some authority with the resources to compute and publicly disseminate the information. Cycles of religious observance, regular market days, and officially proclaimed festivals marked the lapse of multi-day periods, while the time of day or night might be told from a muezzin's call or a church bell, the striking of a gong in the palace or a sundial in the town square (or, towards the end of the medieval period, a mechanical clock).

The ancient "unequal" or "seasonal" hours dividing the periods of daylight and darkness into twelve equal portions apiece persisted throughout most of the medieval period in Europe and the Islamic world, although equal hours were also used. Constant-length fundamental units in other timekeeping systems included the Indian *ghaṭī* or sixtieth of a day and the medieval Chinese *shi* (twelfth of a day) and *ke* (hundredth of a day). Of course, constant-flow devices such as water clocks and sand-clocks marked time units of uniform size (unless constantly recalibrated over the course of the year).

Throughout the medieval world, there were very diverse ways of quantitatively synchronizing various astronomical and social cycles into an annual calendar. Reconciling different calendar systems was sometimes far from simple, as illustrated by such well-known examples as the Christian denominational disputes over how to calculate the date of the Easter festival; the sixth-century Council of Tours decree moving the official new year of the Julian solar calendar from the start of January to March 25; and the sixteenth-century Gregorian calendar reforms that moved it back to January (with additional modifications prompted by the accumulated divergence of Julian calendar dates from actual solar phenomena). The luni-solar calendars of South

Asia and its neighbors were even more complicated, with some versions beginning the year at the (northern hemisphere) spring equinox, or the new moon closest to it, and others at the autumn equinox. Some Indian systems reckoned the month from new moon and others from full moon, and most left it up to individual almanac-makers to determine when an intercalary month was needed to keep lunar and seasonal cycles in sync.

In 632, a Qur'ānic injunction famously abolished intercalation in the Islamic or Hijri calendar, which thereafter was strictly lunar. But the start of a given month in that calendar, being dependent on local observation of the new crescent moon, might vary from one region to another. Meanwhile, other systems such as the ancient Persian solar calendar and the Seleucid calendar were employed alongside the Hijri calendar for administrative purposes in medieval Islamic states, and many of their subjects followed Christian or Jewish (or Sabian, Zoroastrian, etc.) ritual calendars as well. The Chinese luni-solar civil calendar of this period, whose year typically began at the second new moon after the winter solstice, was officially promulgated by imperial administrations in annual almanacs. It, too, coexisted with remnants of ancient agrarian and festival calendars. Moreover, many calendric systems that differed in structure shared some individual features: e.g., the seven-day week cycle that was common to all the calendars (from Iceland to India and beyond) that were influenced by the concepts of ancient Greco-Babylonian astrology.

Keeping track of the routine passage of time in medieval daily life, at least for those who ventured to other regions or countries, thus required a high degree of metrological flexibility. The more complicated refinements of timekeeping involved in calendar-making or astrological prediction were generally considered the province of mathematical professionals.

HOUSEHOLD AND COMMERCIAL NUMERACY

The most common application of everyday numeracy in medieval times, as in antiquity, was the management of the home and the labor and business that supported it: calculating wages, purchases, household produce, and so forth. Throughout the many cultures and periods here subsumed under the term "medieval," references in contemporary literature reveal that women were responsible for much of this quotidian accounting work, although the majority of practitioners of commercial and craft numeracy were men.

In the rural farming households where most people lived, home production including women's labor was frequently an alternative to commercial transactions. One example is the "huswyfe" in a fifteenth-century English ballad who annually wove "a pece off lenyn and wolen . . . For to clothe owre self and owr cheldren . . . Elles we shold go to the market, and by het ful deer." But even feudal peasants regularly participated in the money economy, as indicated in a thirteenth-century French poem's reference to their typical possessions including "borse et aumosniere" (money-bag and purse).

Besides the sex-specific professions of midwife, wet-nurse, and sex worker, women earned income through employment as servants and agricultural laborers, as well as the sale of excess home produce. The practice of rural women bringing their agricultural products to sell in nearby cities in thirteenth- and fourteenth-century Egypt and Tunisia is mirrored in the English Master Fitzherbert's 1534 *Book of Husbandry*:

> It is a wyues occupation . . . to go or ride to the market, to sel butter, chese, mylke, egges, chekyns, capons, hennes, pygges, gese, and all maner of cornes. And also to bye all maner of necessarye thynges belongynge to houssholde, and to make a trewe rekenynge and a-compte to her housbande, what she hath payed. And yf the housbande go to the market, to bye or sell, as they ofte do, he than to shewe his wife in lyke maner.

Similarly, ancient and medieval Sanskrit texts of social/religious law declare that a wife should be prudent in her household and charitable expenditures; she maintains the family savings and purchases goods from traders. If she runs short of money while her husband is absent, she should support herself financially by crafts such as spinning.

Numeracy was even more ubiquitous in the lives of urban communities. More of their household needs were met through purchase rather than direct production, and many earned their living entirely through paid labor. Besides receiving and spending wages, women might have to manage money or other quantitative tasks in their jobs: for instance, as a vendor in a market, or a tax-collector from female vendors at the city gates, or a money-collector at public baths for women. Women of property in particular frequently required numeracy skills, whether they were running an inn in sixth-century Byzantium, buying raw materials for their silk workshop in late medieval London, or making a donation to a temple in thirteenth-century south India.

In medieval China, women sustained cottage industries like the production of textiles, and in *The Mathematical Classic of Sun Zi* there is a problem devoted to a girl skillful at weaving who doubles her weaving over a period of five days. Given the total amount of material produced, the problem is to determine how much she actually produces each day. Such problems appear in the earliest of Chinese mathematical texts, including the Han dynasty *Book on Numbers and Computations*, where exactly the same problem appears with the same total amount of material woven, although the fractions are not reduced to lowest terms as they are in the *Sun Zi* text. The same problem again appears in the *Nine Chapters* (Chapter 3, Problem 4) with exactly the same numbers but with fractions reduced to lowest terms as in the *Sun Zi* version.

Perhaps the most elementary of the medieval records and artifacts related to everyday calculation was the split tally-stick, a length of wood into which notches were carved representing quantities of goods, amount of a debt, etc. The notched stick would be split lengthwise, and one half retained by each of the two parties to the contract, as a record of the agreed-upon amount. For

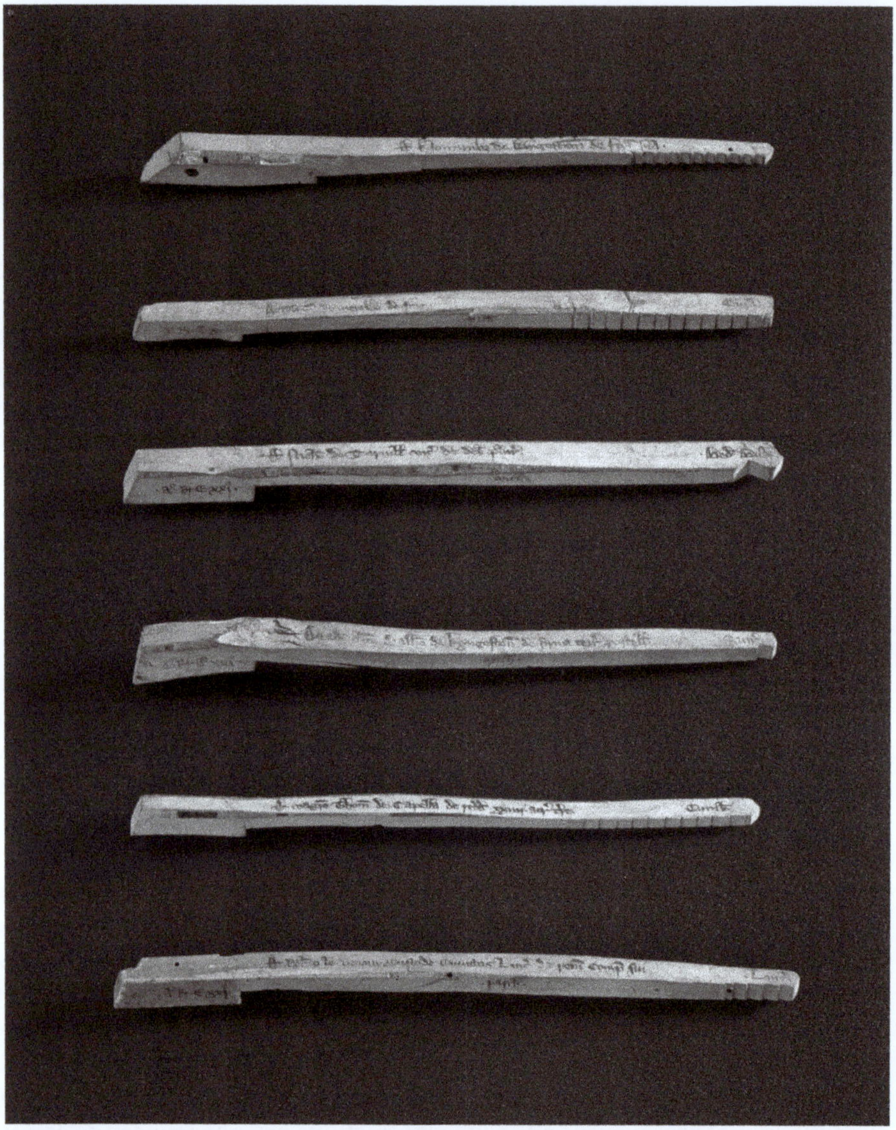

FIGURE 1.1. "Stocks" or creditor's halves of wooden tally sticks from mid-fifteenth-century England, used to document financial transactions. The debtor would hold the matching piece or "foil" of the stick, marked with the same notches, to ensure the fidelity of the record. The British Exchequer continued to use tally sticks until the early nineteenth century. Photo by SSPL / Getty Images.

literate users, quantitative transaction records included account books or ledgers, tax rolls, invoices, receipts, contracts, and many other documents. Such records might mix different number systems and scripts, as in an agricultural account on papyrus from ninth-century Egypt with Arabic personal names accompanying amounts denoted in Greek alphabetic numerals.

Likewise, in medieval China a custom from ancient times was the use of wooden tallies into which notches were carved to keep a record of contracts, deeds, bonds, or obligations of various sorts. Creditors usually kept the left side of a tally, debtors the right half. Even when, in later times, contracts were drawn up on paper, in writing, wooden tallies were still used with notches recording the financial details for later verification upon fulfillment of the contract.

Accounting of course required not only numerate computation but standardized currency systems. In most of Eurasia, many medieval minted coins were offshoots of ancient Greek and Roman currencies, sometimes replacing or coexisting with earlier forms of metal coinage. For example, the Greek silver drachma (or six obols) was an ancestor of Sasanian *drahm* and Islamic *dirham* coins as well as the Indian *dramma*. The Roman silver *denarius* and gold *denarius aureus* (the latter replaced by the Late Roman *solidus* or *bezant*) gave rise to the Islamic *dinar* and Indian *dinara*. The Islamic copper or bronze *fals* was similarly derived from the Roman/Byzantine copper *follis*. In Europe, Carolingian currency reforms based on the *denarius* (*denier*, penny) as one-twelfth of a *solidus* (*sou*, *soldo*)—twenty of which constituted a *libra* (*livre*, *lira*, pound)—shaped the basic structure of various money systems well into the modern period.

But medieval trade and finance worldwide comprised much more than just the exchange of physical coins for material goods. More abstract devices such as mercantile credit accounts, letters of credit, and various forms of investment and partnership facilitated the commercial expansion of the later medieval period. A sample problem from a ninth-century south Indian mathematics textbook in Sanskrit gives an idea of the financial sophistication of such practices:

> Four merchants who had invested their money in common were asked each separately by the customs officer what the value of the commodity was; and one eminent merchant, deducting his own investment, said that that was in fact 22. Then another said that it was 23; then another said 24; and the fourth said that it was 27; each of them deducted his own invested amount. Friend, tell me separately the value of the commodity owned by each. (Rangacarya 1912: 136–7)

Craft mathematics, especially among the non-literate, encompassed many aspects of elementary arithmetic and geometry, existing more or less on the fringes of formal academic mathematics. Our knowledge of this activity comes largely from the traces it left in academic mathematics, such as a pamphlet on

FIGURE 1.2. Measuring weights with scales, in a miniature painting in an eleventh-century Italian manuscript of the ninth-century Latin encyclopedia *On the Natures of Things* by Rabanus Maurus of Mainz. Rabanus' text claims that Moses and Pheidon of Argos were the first to know the science of weight; the figure on the left holding a balance (*trutina, libra*) may represent the former, with the latter on the right holding a steelyard (*statera*). Photo by DeAgostini / Getty Images.

the geometry necessary for constructing decorative tiling patterns written by the tenth-century Baghdad mathematician Abū al-Wafā' al-Būzjānī. Abū al-Wafā' complained that craftsmen making such patterns use constructions without proofs and don't distinguish between exact and approximate constructions. Similar formalizations of practical empirical rules exist in sections of Sanskrit arithmetic treatises devoted to the measurement of sawn lumber, heaps of grain, and excavations, as well as in the surveying manuals of medieval Europe.

As formal education became more widespread (though nowhere near universal) in the medieval world, elementary mathematical curricula gradually evolved. In China, a compilation of mathematical works later known as *The Ten Mathematical Classics* was studied by candidates for imperial administrative examinations beginning in the seventh century. Primary schooling or tutoring in medieval Islam might extend to girls as well as boys and also to Christians, Jews, and other non-Muslims. Its curriculum prioritized Qur'ān recitation (although, as one authority advised, not in the case of Christian students) and Arabic literacy and grammar. Mathematical subjects that might be taught included first and foremost the basics of calculation, frequently "Indian calculation" or decimal place-value arithmetic, and perhaps geometry formulas and *arithmetika* or properties of number. Pupils pursuing advanced education would go on to algebra, astronomy, and other more esoteric subjects.

The Greek and Latin churches of emergent Christendom took over parts of the pedagogical and scholarly missions of the pagan academies and libraries, which were largely defunct by the middle of the first millennium. Most crucially, a 793 edict of Charlemagne decreed that all monasteries and cathedrals should include a school for young literate boys, whose elementary education included some rudiments of arithmetic and geometry.

On the Indian subcontinent, various ancient and medieval sources, among them Buddhist and Jain canons, preserved some information on the state and extent of general education, including numeracy. The famous Sanskrit manual on statecraft, the *Arthaśāstra* composed by Kauṭilya around the start of the Common Era, mentions that learning the alphabet (*lipi*) was paired with basic arithmetic (*saṃkhyāna*). It also advises future rulers that they should become familiar with various metrological systems and conversion procedures, as well as devoting a part of their schedule each day to "the accounts of receipts and expenditure," to avoid being taken advantage of in financial affairs.

FOLK AND RECREATIONAL MATHEMATICS

The quantitative knowledge applied in the everyday mathematics described in the preceding sections was transmitted not just in formal educational settings, but also in folklore and play. Much of the documentary evidence of such activities (counting rhymes, numerical riddles and games, etc.) dates from late medieval or even post-medieval times: often we can only guess how far their earliest forms might pre-date their surviving records. But the circulation of different versions of the same material in different languages and cultures, often in the form of mnemonic verse, indicates the utility and tenacity of this sort of "folk mathematics."

One example is the familiar sequence of rhymed couplets on the lengths of the months of the (Julian) calendar, appearing in French sources perhaps as early as the thirteenth century and in an (immediately recognizable) English equivalent from the sixteenth century:

En avril, en juing, en septembre	Thirty dayes hath Nouember
A .xxx. jours et en novembre:	Aprill, Iune and September.
Tout li autre ont .xxxj. jour,	February hath xxviij alone,
Fors fevriers qi est li plus cour	And all the rest have xxxi.
En soi que .xxviij. jors n'a . . .	

Besides simple numerical data of the "thirty days" sort, logic and computation puzzles were also transmitted across much of the medieval world. Their history is often tangled and obscure, as in the well-known genre of "river crossing problems": e.g., a goat, a wolf, and a cabbage must be transported across a river in a vessel that will hold only one of them, without ever leaving the wolf alone with the goat or the goat with the cabbage. Such puzzles abound in many variant forms requiring the solver to avoid throwing together, for example, unmarried couples or pairs of enemies. Attested in sources from Europe, Africa, and elsewhere, they may have been sometimes transmitted and sometimes independently reinvented.

The famous mathematical riddle that became known as the "hundred fowls problem" (an indeterminate problem involving two equations in three unknowns) strikingly illustrates the cross-cultural popular appeal of even computationally challenging puzzles. A mid-seventh century Chinese version explains it as follows:

> Now one cock is worth 5 *qian*, one hen 3 *qian* and 3 chicks 1 *qian*. It is required to buy 100 fowls with 100 *qian*. In each case, find the number of cocks, hens and chicks bought.
>
> Answer says: 4 cocks worth 20 *qian*, 18 hens worth 54 *qian*; 78 chicks worth 26 *qian*.
>
> Another answer: 8 cocks worth 40 *qian*, 11 hens worth 33 *qian*, 81 chicks worth 27 *qian*.
>
> Another answer: 12 cocks worth 60 *qian*, 4 hens worth 12 *qian*, 84 chicks worth 28 *qian*.
>
> Method says: Add 4 to the number of cocks, subtract 7 from the number of hens and add 3 to the number of chicks to obtain the answer. (Lam 1997: 236)

In the Latin version recorded by the eighth-century English scholar Alcuin of York:

> A merchant and 100 pence. A merchant wanted to buy a hundred pigs for one hundred pence. For a boar he would pay 10 pence; and for a sow 5 pence; while he would pay one penny for a couple of piglets. How many boars, sows and piglets must there have been for him to have paid exactly 100 pence for the 100 animals?
>
> Solution. Take 9 sows and a boar, to the value of 55 pence altogether, and 80 piglets at 40 pence, which brings the total of pigs to 90. For the other 5 pence take 10 piglets, and that brings the total of pigs and of pence to 100. (Hadley and Singmaster 1992: 106)

And as described in a ninth-century Sanskrit textbook by the mathematician Mahāvīra:

> Pigeons are sold at the rate of 5 for 3 (*paṇas*), *sārasa*-birds at the rate of 7 for 5 (*paṇas*), swans at the rate of 9 for 7 (*paṇas*), and peacocks at the rate of 3 for 9 (*paṇas*). A certain man was told to bring at these rates 100 birds for 100 *paṇas* for the amusement of the king's son and was sent to do so. What (amount) does he give for each (of the various kinds of birds he buys)? (Rangacarya 1912: 134–5)

Although the livestock and currency denominations in different instances of this riddle are adjusted for their various local audiences, they preserve the memorable key theme of purchasing some combination of a hundred creatures for a hundred coins. The open-endedness of the problem and the trial-and-error experimentation needed to solve it may also have contributed to its widespread popularity. (Formal indeterminate-equation techniques for analytical solution of such a problem would have been beyond the capability of general audiences.)

Basic numeracy techniques such as counting, analysis, logic, and the application of simple algorithms were combined with manipulation of physical objects to create board, card, and dice games. Games of chance in general, based upon the randomization of artificially restricted sets of outcomes, required at least informal understanding of basic probabilistic concepts that were not studied analytically until the early modern period.

The ancient pastime of dice playing attested in many prehistoric societies (where it was apparently connected to divination techniques) gave rise in the ancient Mediterranean world to numerous "tables games." Here, dice rolls were translated into spatial form by moving a counter the corresponding number of steps on a marked board. Medieval European gamers, for example, played variants of the classical *tabula* or "tables," an ancestor of early modern backgammon. Meanwhile in China, numerous games used dice rolls to determine selections from a set of tokens prescribing some form of gameplay. These tokens, printed on paper or carved in some other material, evolved respectively into playing cards (which found their way to Europe by the late fourteenth century) and into the marked tiles of games such as dominoes and modern mahjong.

Games of strategy, reducing or eliminating the element of chance, were another popular form of numerate recreation in the medieval world. The earliest examples include ancient Chinese *Wei chi* with its grid of black and white stones, from which medieval Japanese *Go* is descended, and various mancala games of West Asia and Africa. The best-known medieval strategy game is doubtless the "game of kings" or chess, which originated in India around the sixth century as *caturāṅga*, "quadripartite," referring to the four military forces: elephants, cavalry, chariots, and infantry (corresponding respectively to the modern pieces bishop, knight, rook, and pawn). By the

FIGURE 1.3. Women playing backgammon, 1283. Miniature from *Libro de los juegos* (The Book of Games) by Alfonso X, called The Wise (1221–84). Courtesy of the Royal Library of the Monastery of El Escorial, Spain. Photo by Leemage / Corbis via Getty Images.

seventh century, variant forms called *chatranj* or *shatranj* had been adopted in Persian- and Arabic-speaking regions. Other versions evolved by the end of the first millennium into European "chess" (*xadrez*, *scacchi*, *échecs*, *Schach*, etc.), Chinese *xiangqi*, Japanese *shogi*, and many more.

The 1283 *Book of Games* commissioned by Alfonso X in Toledo contains descriptions of three types of games—chess, dice, and tables—classified as depending respectively on skill, chance, and both skill and chance. Their countless hybrid forms included checkers with dice, decimal chess with dice, four seasons chess, nine men's morris with or without dice, and astrological checkers for seven players. Illustrations in manuscripts of the *Book of Games* testify to the range of participants enjoying game play, from women playing a tables game to a chess contest between a Jew and a Muslim. They also indicate the arithmetic competence and algorithmic finesse required of the players and the enjoyment and recreation that these numerate skills provided.

THE DECIMAL PLACE-VALUE NUMBER SYSTEM

The final aspect of medieval everyday mathematics that we will explore—the evolution and diffusion of the base-10 place-value numerals now used throughout the world—strikingly illustrates the continuing globalization of mathematical knowledge in this period. Although the decimal numerals' initial creation has left no detectable record, their development and subsequent dispersion were probably influenced by various aspects of ancient numeracy, some of which have been described in more detail in Volume 1 of this series and in the Introduction of this volume. All of the following concepts were in existence by the beginning of the Common Era:

- The Mesopotamian base-60 positional system, including representation of fractions as powers of 1/60 and a zero-placeholder symbol. This system survived in the sexagesimal time and angle measures of Hellenistic astronomy written with Greek alphabetic decimal numerals.
- The quasi-place-value decimal Chinese written numbers, and their fully place-value decimal equivalents (including decimal fractions) used in counting-rod computations.
- Engraved "abacus" boards from Greece and Rome, and similar devices elsewhere in the ancient world, whose counting-pebbles (*psephoi*, *calculi*, etc.) enabled place-value calculation without written numerals.
- The (non-place-value) system of glyphs for decimal numbers in the left-to-right Brahmi script of late-BCE South Asia.
- Ancient orally transmitted Sanskrit texts containing verbal allusions to names for large powers of ten and also to a symbol that was attested centuries afterward as the standard glyph for zero (Sanskrit *śūnya*).

Indian Decimal Numerals in South and East Asia

Sometime within the first few centuries of the Common Era, a fully developed decimal place-value system for integers became firmly established in Indian texts on astronomy and mathematics. Unfortunately, the earliest stages of this system's evolution have proved stubbornly resistant to a sharp chronological focus. We do know that Indian astronomers by the early centuries CE were familiar with the Brahmi non-positional system of numeral glyphs including the individual digits 1 through 9 and almost certainly also with a symbol in the form of a circle or thick dot for expressing *śūnya*, "emptiness" or the absence of a number. Perhaps some of their ideas may also have been influenced by the efficient versatility of the counting rods used by Chinese traders or pilgrims, or even by base-60 positional numbers with zero possibly encountered in Indo-Greek astronomy.

Moreover, some contemporary philosophical works clearly reveal familiarity with practical examples of positional reckoning. For example, the fourth- or fifth-century Buddhist scholar Vasubandhu attributed to his second-century predecessor Vasumitra a simile involving the changing significance of the same

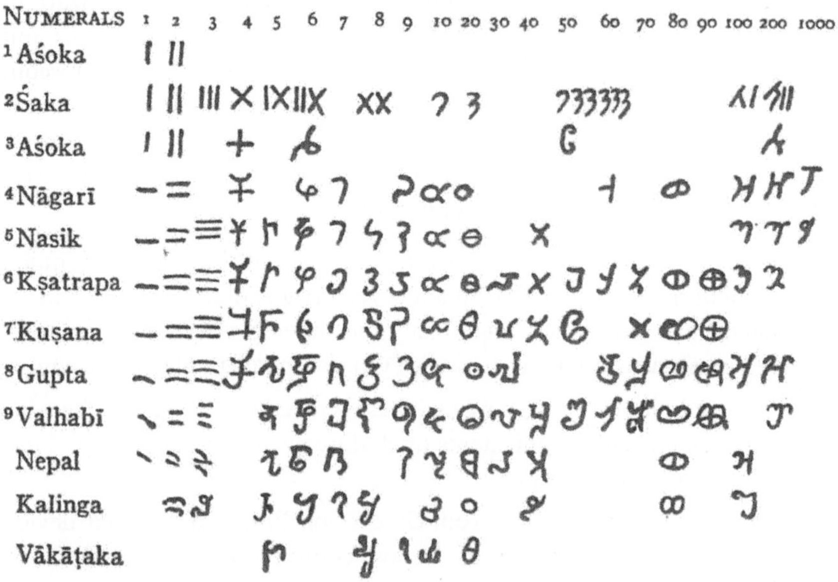

FIGURE 1.4. Approximately chronologically ordered samples of non-positional number glyphs used in inscriptions from early India. The (mostly) ciphered-additive numerals of the ancient Brahmi script are shown in the third through about eighth rows (Smith 1911: 25). Courtesy of University of California Libraries / Internet Archive.

counting-token when placed in the units position, the hundreds position, etc. A commentary on the famous *Yogasūtra*, perhaps from the same period or a little later, employs a similar comparison between the different meanings of one line marked in the hundreds or tens or units place and the different designations of one woman as mother or daughter or sister.

The existence of a fully mature version of the ciphered-positional decimal system is attested in a Sanskrit treatise composed about 500 by the astronomer-mathematician Āryabhaṭa. This work explicitly describes the place-value principle and positional techniques for extracting square and cube roots of decimal integers. Subsequently, the sixth and seventh centuries supply the first securely dated evidence of two systems for verbally encoding decimal place-value numerals by assigning to each of the ten digits various Sanskrit words or letters. One of these is the so-called "word-numeral" system where, for example, the word "moon" means 1, the words "hand" and "eye" both signify 2, "tooth" means 32, and so on through a large set of conventional synonyms for the single-digit numbers and several two-digit ones. The other is an alphabetic system in which certain consonants of the Sanskrit alphabet stand for 1, others stand for 2, and so on, while vowels have no numerical meaning. Multi-digit numbers could thus be easily represented in either system by strings of actual

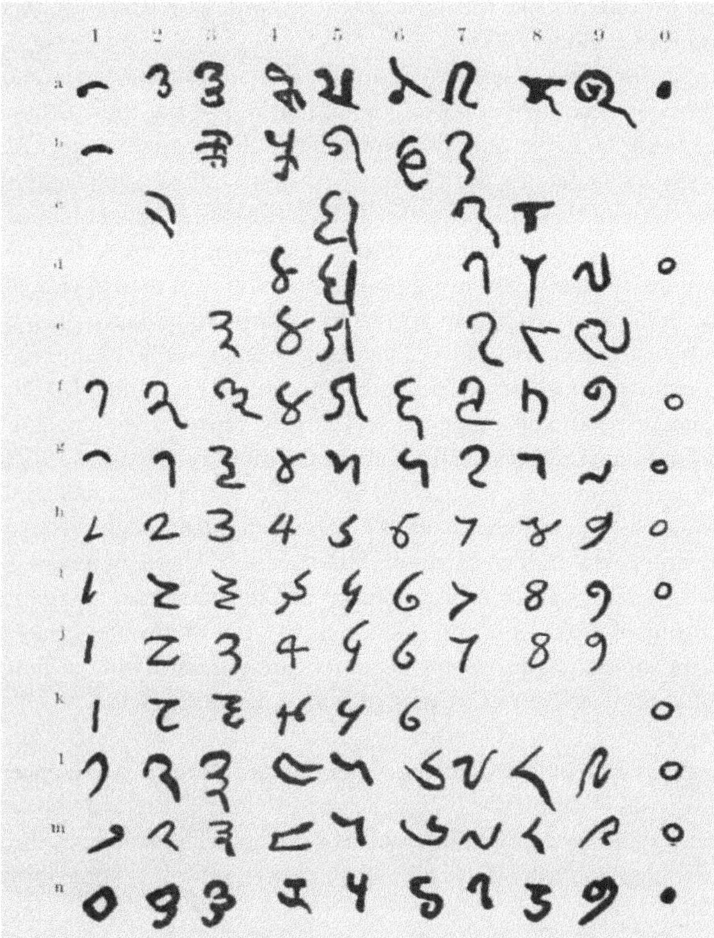

FIGURE 1.5. Samples of place-value decimal (ciphered-positional) number glyphs used in Indian inscriptions and manuscripts from about the mid-first to mid-second millennium (Smith 1911: 49). Courtesy of University of California Libraries / Internet Archive.

Sanskrit words. The ordering of a number's digits in both systems invariably proceeded from the least significant to the most significant, the reverse of their written numerical order in left-to-right Indic scripts. For instance, the word-numeral expression "hand-month" would be verbally equivalent to "two-twelve" but numerically equal to 122.

Neither astronomical nor arithmetical works in Sanskrit during this period used decimal fractions: they represented arbitrary fractions as an integer numerator placed above an integer denominator, the ancestral format of our modern "common fractions." There seems to have been no particular preference in Indian arithmetic for the unit fractions widely favored in the ancient Mediterranean region.

Epigraphic evidence suggests (cautiously, since inscriptional styles tend to be conservative or even deliberately archaic) that use of the Indic decimal place-value numerals, and the Sanskrit word-numeral system for expressing them verbally, spread widely in the "Sanskrit cosmopolis" of South and Southeast Asia in the sixth and seventh centuries. In fact, the earliest unambiguous physical record of the decimal place-value zero is, famously, the recently rediscovered inscription near Angkor Wat in Cambodia mentioning a date in year 605 of the Śaka era, or 683 CE. A hybrid format is seen in some early exemplars: for example, an inscription around 578 denoting the number 80 by the combination of a non-place-value glyph for 80 with a (technically redundant) zero symbol. Fully ciphered-positional numerals had largely replaced the ancient additive numerals in Indian inscriptions from about the ninth century onwards, although various other number systems survived for centuries in more limited applications.

Knowledge of the Indian numerals in China apparently dates from about the eighth century via Indian astronomical sources. These numbers remained primarily a specialized "scientific notation" in Chinese texts relating to Sanskrit-derived forms of astronomy and calendrics. Even in that context they were not particularly popular: as an eleventh-century Chinese history of the Tang dynasty remarked about a calendrics treatise of Indian origin:

> The computation of [the Catching Nines Calendar] used pen and paper and not counting rods. The method is complicated and fragmented and you are lucky if you get it right. It cannot be adopted as a method. The writing of the numbers is very strange and from the first impressions it is difficult to argue for it. (Li and Du 1987: 108)

The Chinese use of zero distinguished between the two technical terms 空 *kong* (empty, vacant), used in astronomical tables, and 零 *ling* (zero), pertaining to mathematical works.

"Indian Numerals" in West Asia and the Islamic World

The westward spread of the decimal numerals is first attested around the middle of the seventh century, in the Syrian bishop Severus Sebokht's famous praise of the "nine signs" of the Indians' "subtle inventions in this teaching of astronomy." (The phrase "nine signs" was probably a conventional expression implying not an incomplete set of numerals without zero, but rather a distinction between "signs" for nonzero digits and the zero symbol standing for the absence of a digit.) Severus, himself an astronomer, was especially impressed with the "reckoning that goes beyond words" in this "rational method of calculation"—

meaning, presumably, its capacity to handle multi-digit numbers too large to state in the Greek or Syriac mathematical vocabulary of the time.

Exactly how the Indian numerals migrated to West Asian centers of learning such as Sebokht's Qenneshre in Syria and later the 'Abbāsid caliph's court in Baghdad is not clearly understood. In the early to mid-first millennium the Sasanian city of Jundishapur in what is now southern Iran was renowned for its scholarship not only in Pahlavi, Syriac, and Greek but also in Indian works, among others. Indian astronomy and its decimal place-value numbers may have been known there.

Moreover, Indian scientists who visited Baghdad in the late eighth century likely demonstrated these numbers to their Arabic-literate counterparts, with accompanying arithmetic techniques that became known as "Indian calculation" (*ḥisāb al-hind*) or "dust calculation" (*ḥisāb al-ghubār*). It was in Baghdad in the early ninth century that the astronomer-mathematician Muḥammad ibn Mūsa al-Khwārizmī wrote the first known Arabic textbook on this form of calculation, which now survives only in twelfth-century Latin versions.

The earliest work on decimal arithmetic still extant in Arabic is the 952 *Indian Calculation* of al-Uqlīdisī of Damascus. (His name means "the Euclidean" or copyist of the *Elements* in Arabic, which formed part of the professional activities of several noted Muslim mathematicians.) Al-Uqlīdisī and other Arabic authors wrote the "Indian numerals" with the highest digit on the left, as they were originally written in Indian texts and as we still write them today. They called the zero-sign *ṣifr*, derived from a Semitic root meaning "empty," and apparently translating (or coinciding with) the sense of Sanskrit *śūnya*.

According to al-Uqlīdisī, some calculators avoided this arithmetic out of distaste for the dust-board on which the figures were scratched and erased: it dirtied the hands, it was liable to smearing, it was unfavorably associated with the marketplace scribe or the street astrologer, etc. His proposed solutions included modifying dust-board arithmetic techniques to adapt them to the paper writing medium rapidly diffusing through the Islamic world and substituting Arabic or Byzantine Greek alphabetic numerals for the recognizably "Indian" decimal digits.

Al-Uqlīdisī also explained how to carry out in "Indian calculation" various computations typically performed by finger-reckoning, with non-positional numerals, or with the astronomers' base-60 numbers. His comprehensive decimal arithmetic included some of the earliest known procedures for manipulating decimal fractions, which were expounded more fully in works by later mathematicians such as al-Samaw'al (twelfth century) and al-Kāshī (fifteenth century). More typically, common fractions in decimal numerals were represented using the Indian vertical format, which by the twelfth century was modified in the Islamic West by the now-familiar horizontal bar separating numerator and denominator.

The decimal place-value Indian numerals, although highly appreciated by mathematicians in the Islamic world, did not fully displace rival systems of arithmetic in practical applications. Various forms of business arithmetic continued to favor mental calculation and finger-reckoning, and fractional quantities were still conventionally written as sums of unit fractions. And the traditional "astronomers' calculation" with base-60 numbers recorded in alphabetic *abjad* digits still prevailed in astronomical literature, including the Arabic and Persian works introduced to Chinese astronomers during the twelfth-century Mongol conquests. Despite being assiduously studied and translated in scholarly circles of the Yuan and Ming dynasties, such works do not seem to have had any impact on Chinese authors' use of their own ancient form of decimal place-value numeration.

The circulation of decimal place-value arithmetic in Islamic mathematics was nonetheless active enough to introduce regional variations in the shape of the numerals and in related notation. In particular, the original "nine signs" of the *ghubār* or dust numerals evolved into two main families of glyphs, one typically used in the Mashriq or eastern part of the Islamic world, and the other specific to the Maghrib or western regions. Speculation on the precise causes of the divergences in their forms is ongoing but thus far inconclusive.

"Indian" and "Arabic" Numerals in the Latin West and Byzantium

The first known examples of decimal place-value numerals in Latin occur in late tenth-century manuscripts from Visigothic Spain, one of which includes the accompanying encomium:

> We must know that the Indians have a most subtle talent and all other races yield to them in arithmetic and geometry and the other liberal arts. And this is clear in the nine figures with which they are able to designate each and every degree of each order (of numbers). (Burnett 2006: 17)

Latin works typically use the Western forms of the numeral glyphs, which ultimately evolved into the shapes of the modern "Indo-Arabic" numerals, including the zero-symbol termed *chifra* from Arabic *ṣifr* (also "cifra," "zephir," etc. whence "cypher" or "cipher" in English and cognates such as "chiffre" or "Ziffer" in other Western European vernaculars). Its name was sometimes applied to the decimal digits in general, an ambiguity that still persists in idioms like "ciphering" to mean calculating with written figures and "a mere cipher" to denote something of zero importance.

In the eleventh and twelfth centuries, writers in Latin grew increasingly familiar with these novel figures and their associated arithmetic via numerous translations and expositions produced in places like Toledo and Antioch. Some Latin scholars with an interest in the subject studied Arabic to read the source texts directly, some Latinized the vernacular explanations of Muslim or Jewish intermediaries, and some merely disseminated or modified previous Latin descriptions. A few early attestations betray some initial confusion about the ordering of the place-value digits, as the difference in writing direction between Arabic and Latin scripts suggested that the numerical notation should flip direction, too.

The "decimalization" of abacus arithmetic attributed to the tenth-century bishop and translator Gerbert of Aurillac doubtless reinforced the convention of writing the highest digit on the left. Gerbert introduced the use of markers or "apices," each marked with a decimal digit, to represent place-value numbers in the abacus columns with units on the right. (The ease with which the markers could be rotated probably contributed to the diversity of orientations in which some of the individual digits were written before attaining a standardized form.) The zero-symbol was sometimes omitted from this set of figures as it was not needed to designate an empty place on the abacus with fixed columns; when included it was generally called the *rotula*, "little wheel," from its circular shape.

An early name for this foreign arithmetic, "helcep sarracenicum" ("*al-ḥisāb* of the Saracens"), was generally superseded by another neologism derived from a translator's opening phrase "Dixit Algorizmi" ("al-Khwārizmī says") in a twelfth-century Latin recension of al-Khwārizmī's now-lost ninth-century Arabic treatise. The name "Algorizmi," "Alchorismi," etc. morphed into a technical term for these computational techniques themselves: "algorismus" or "algorism," the ancestor of our modern word "algorithm." The abacus or "Pythagorean table" with its board and counters, and the algorism with its written ciphers, became the two chief methods for practical calculation in Europe's late middle ages.

However, to many contemporaries of the early thirteenth-century mathematician Leonardo of Pisa (Fibonacci), "abaco" (or "abbaco," "abbacho," etc.) denoted computation in general rather than the physical abacus or counting-board. Leonardo intended his 1202 work *Book of Calculating* (*Liber abaci*) to demonstrate the superiority of the "method of the Indians" over not only the counting-board but also the algorism techniques currently in use:

> [F]ollowing my introduction, as a consequence of marvelous instruction in the art, to the nine digits of the Indians, the knowledge of the art very much appealed to me before all others . . . [A]ll this, the algorism as well as the Pythagorean arcs [grouping columns on the counting-board], I considered as almost a mistake compared to the method of the Indians . . .
>
> The nine Indian figures are: 9 8 7 6 5 4 3 2 1. With these nine figures, and with the sign 0 which the Arabs call zephir, any number whatsoever is written . . . (Sigler 2002: 15–16)

It is not clear in exactly what ways Leonardo considered his "modus Indorum" to be fundamentally different from the decimal reckoning of the counting-board or early algorism texts, but his treatment of it was distinguished by its clarity and comprehensiveness and it proved highly influential.

The complicated business arithmetic developed in the expansion of international commerce after the twelfth century led to increasing reliance on decimal place-value arithmetic, despite the notorious distrust of it expressed by some merchant guilds and other bodies. The so-called "abacus schools" of Italy provided training in its techniques (not in the use of the counting-board, as the name might suggest), as well as other methods of practical computation. Their teachings soon diffused into other European communities and vernacular languages. The bar-separated vertical notation for common fractions and their arithmetic rules spread along with the arithmetic of decimal integers; decimal fractions were a later (re-)invention in the Latin corpus.

The late medieval period saw dissemination of the "Indian numerals" into languages of Eastern Europe. Hebrew and Greek mathematical texts sometimes combined the place-value principle with letters of the Hebrew, Greek, or Persian alphabet, as previously suggested by al-Uqlīdisī. Both western and eastern forms of the "cipher" glyphs appeared in some Byzantine Greek manuscripts (as did the decimal fraction notation of al-Kāshī). The first Hebrew work to adopt them was probably that of Mordecai Comtino in the mid-fifteenth century in Constantinople.

As these number signs became naturalized in European pedagogical texts, their early associations with the "wisdom of the Indians" became embellished with pedantic speculations and outright fables. For example, according to an apocryphal story repeated by the twelfth-century Spanish Jewish philosopher and mathematician Abraham ibn Ezra, it was a Jewish envoy to India from the first ʿAbbāsid caliph al-Ṣaffāh around 750 who persuaded an Indian sage to introduce the decimal system in the West. Ibn Ezra was presumably influenced by his knowledge of the many Jewish merchants of his own day active in the India trade:

> [F]or a large sum, one of the wise men of Arin [the Arabized name of Ujjain in Madhya Pradesh] agreed to go to the king. The Jew swore to him that he would detain him for only one year and that he would return him to his home. The scholar, whose name was Kanaka, was brought to the king, and he taught the Arabs the basis of numbers, i.e., the nine numerals. (Pingree 1968: 101)

The thirteenth-century scholar John of Sacrobosco in his *Algorismus Vulgaris* or *Popular Arithmetic* reasoned instead that the right-to-left direction of Semitic scripts was the source of the right-to-left ascending order of the decimal digits:

> In this art we write [numbers] in the Arabic or Judaic way of the inventors of this science. (Burnett 2009: 7)

In a fourteenth-century English arithmetic treatise we can see traces of both al-Khwārizmī and the "wise man of Arin" in the new character of an Indian "King Algor," along with an alternative and equally fanciful Greek etymology:

> And þis boke tretys þe Craft of Nombryng, þe quych crafte is called also Algorym. Ther was a kyng of Inde, þe quich heyth Algor, & he made þis craft. And after his name he called hit algorym; or els anoþer cause is quy it is called Algorym, for þe latyn word of hit s. Algorismus comes of Algos, grece, quid est ars, latine, craft on englis, and rides, quid est numerus, latine, A nombur on englys . . . Þis present craft ys called Algorismus, in þe quych we vse teen signys of Inde. Questio. Why ten fyguris of Inde? Solucio. for as I haue sayd afore þai were fonde fyrst in Inde of a kynge of þat Cuntre, þat was called Algor. (Steele 1922: 3)

> [And this book treats the craft of numbering, which craft is called also Algorym. There was a king of Inde, who was called Algor, & he made this craft. And after his name he called it algorym; or else, another cause is why it is called Algorym, for the Latin word of it, Algorismus, comes from "Algos" in Greek, which is "ars" in Latin, "craft" in English, and "rides," which is "number" in Latin, "a number" in English . . . This present craft is called Algorismus, in which we use ten signs of India. Question: Why ten figures of India? Answer: For as I have said before, they were found first in India by a king of that country, that was called Algor.]

CHAPTER TWO

Practice and Profession

The Teacher, the Trader, the Almanac-Maker: Livelihoods of Medieval Mathematics

CLEMENCY MONTELLE AND KIM PLOFKER

Besides the everyday numerate practices surveyed in Chapter 1, the functioning of the medieval world also relied on more esoteric systems of mathematical knowledge developed and applied by trained specialists. The work of these specialists was of course very different from modern conceptions of mathematical "professionalization" and the job title "mathematician," which are largely a product of the nineteenth century. Nonetheless, courts and monasteries, observatories and family temples, building sites and marketplaces all required the services of practitioners of various mathematical skills. They might be supported by royal or private patronage, government salaries, returns from business enterprises, income from teaching, or fees for specific services such as making horoscopes or almanacs. The following two accounts provide a glimpse into some of the many diverse lives of medieval numeracy professionals.

On October 13, 1282, in the thriving Italian commercial city of Siena, one Ugo Gigone contracted with the great Salimbeni merchant company (*societas*) to serve as their factor or agent for the subsequent four years. The Salimbeni family around this time was at the height of its financial and political power, with extensive lands and tax collecting and banking interests as well as overseas trading in spices, silks, and other commodities. As a representative of his company in conducting transactions and negotiations, a factor needed the standard numerate skills of commerce and accounting, while as a salaried employee he was exposed to temptations to exploit the company's resources for his own advantage. Ugo pledged:

> to go and to remain wherever you wish and order me throughout Tuscany and Lombardy and the kingdoms of France, England and Sicily and elsewhere, wherever you direct and wish to do business and to gain a profit and to carry on your business well and advantageously and lawfully . . . And to return and to consign to you or to any one of you or to whomever you wish . . . a correct and legal accounting of each and all . . .
>
> And I also promise you that whatever may be donated or given to me in money or gold or silver or anything else by any person or locality or community or baron or prelate in any way, so long as I remain your factor, I shall turn over and give all to you . . .
>
> And I promise you that as long as I remain your factor, as is stated above, I shall not gamble in any game of dice with pledge or money, nor shall I have carnal relations with any married woman, virgin, or religious, nor shall I make any expenditures on them out of your goods . . .
>
> Furthermore I promise you not to make or to form any *societas* with any person or persons without your permission and will. I assert and acknowledge that I do not have and did not invest in your *societas* or in another *societas* any amount of money or any cash . . .
>
> And I promise you to undertake and to observe all these aforesaid conditions, article by article, under penalty of 100 marks of silver, which I promise to give you, just as is stated, if I do not observe or if I act contrary to the conditions . . .
>
> And I do and I promise this for you because you have promised and made agreement to give me 450 Sienese pounds as my fee and salary for the said four years . . . (Lopez and Raymond 1967: 215–17)

The size of the penalty decreed for violation of the contract—at his salary rate, Ugo would have forfeited several months' pay—testifies to the financial power held by a factor in his employer's enterprise.

In Ugo's "correct and legal accounting" of all his transactions, currency exchanges would have figured largely. The original Western European monetary structure—twelve *denarius* in a *solidus*, twenty *solidus* in a *libra*—in practice fragmented into a bewildering variety of exchange rates. A rule-of-three currency conversion problem in Leonardo of Pisa's 1202 *Book of Calculating* illustrates some of the required (decimal place-value) numerate skills:

> If one *soldus* of imperials, which is 12 *deniers* . . . is sold for 31 *deniers* Pisan, and it is asked how many *deniers* Pisan anyone should obtain for 11 imperial,

> you write out the problem. That is, first . . . 12 *deniers* [imperial], then on the same line to the other side you write its value, namely 31 *deniers* Pisan, and place 11 imperial under 12 imperial, as is shown here. And you will multiply the numbers on the opposite sides, that is, 11 by 31; it will make 341. And divide this by 12, [the result] will be 5/12 28 *deniers*. (Lopez and Raymond 1967: 343)

Nearly 140 years after Ugo signed his employment contract, another mathematical professional took up a rather different job. The Iranian scholar Jamshīd al-Kāshī, already a noted mathematician and astronomer at about 40 years of age, arrived in Samarkand in 1421 at the invitation of the Timurid sultan Ulugh Beg, who was himself skilled in mathematical sciences as well as a generous patron of them. Al-Kāshī wrote long letters to his father back home, recording invaluable details of the professional activities of scholars and students at the Samarkand observatory: the numbers of students and teachers, their stipends, their topics of study, their "office politics," and much other information.

> This art (i.e., mathematics) is taught at twelve places . . . There are twenty-four calculators (*mustakhrij*), some of whom are also astronomers and some have begun (studying) Euclid . . . Further, His Royal Majesty (i.e., Ulugh Beg) visits the school once every few days, and attends a class for some time, and I am in His Majesty's company. (On these occasions) we do not know at all what the lesson will be about, nor what problem will be brought up, (whereas) the teacher and students have studied it together the night before . . .
>
> His Royal Majesty had donated a charitable gift amounting to thirty thousand *kopakī* dinars (a type of Iranian gold coin), of which ten thousand had been ordered to be given to students . . . ten thousand-odd students steadily engaged in learning and teaching, and qualifying for financial aid, were listed (as recipients). There are the same number (of students) among the notables and their sons, who dwell in their own homes . . .
>
> . . . when I arrived here, as it is customary, all of them (i.e., the calculators) began inquiring and spying in order to find out the extent of my knowledge in this art (i.e., mathematics), so that, if they found me to have little knowledge or to be equal or slightly superior to them, they keep their positions; thus they might not (have to) say that a stranger got the better of them . . . I solved all (those problems), and submitted (my solutions) to His August Majesty. (Bagheri 1997: 243–46)

EXPANSION OF MATHEMATICAL PROFESSIONS

As nations grew in the early medieval era, so, too, did large-scale aspirations for trade, enterprise, and infrastructure. The fulfillment of these aims, as well as the management of increasingly complicated timekeeping models and astrological predictions, depended on an expanding (middle) class of educated professionals with specialized mathematical skills. Many of their surviving writings reveal an increasingly confident affirmation of the value and coherence of their discipline, a sense of computational practice as an elite unified body of knowledge.

In China around 400, for example, the *Mathematical Classic of Master Sun* declared:

> Mathematics [governs] the length and breadth of the heavens and the earth; [affects] the lives of all creatures; [forms] the alpha and omega of the five constant virtues (i.e. benevolence, righteousness, propriety, knowledge, and sincerity); [acts as] the parents for *yin* and *yang*; establishes the symbols for the stars and the constellations; [manifests] the dimensions of the three luminous bodies [i.e., sun, moon, and stars]; maintains the balance of the five phases [i.e., metal, wood, water, fire, and earth]; [regulates] the beginning and the end of the four seasons; [formulates] the origins of myriad things; and [determines] the principles of the six arts. (Lam and Ang 2004: 189)

This did not always imply superior status for mathematics vis-à-vis other branches of learning, however. As the sixth-century scholar Yan Zhitui wrote in his *Family Instructions for the Yan Clan*:

> Mathematics is an important subject in the six arts [rituals, music, archery, charioteering, calligraphy, and mathematics]. Through the ages all scholars who have participated in discussions on astronomy and calendars have to master it. However, it is a minor occupation, not a major one. (Martzloff 2006: 76)

A textbook by the Indian mathematician Mahāvīra in the ninth century expresses the same sort of broadly eulogistic spirit earlier shown by Master Sun:

> In all those transactions which relate to worldly, Vedic, or other similarly religious affairs, calculation is of use. In the science of love, in the science of wealth, in music and in the drama, in the art of cooking, and similarly in medicine and in things like the knowledge of architecture. In prosody, in poetics, in logic and grammar and such other things, and in relation to all that constitutes the peculiar value of [all] the [various] arts: the science of

computation is held in high esteem. In relation to the movements of the sun and other heavenly bodies, in connection with eclipses and the conjunctions of planets, and in connection with the "three questions" [i.e., geographical location, direction and time] and the course of the moon – indeed in all these [connections] it is utilized . . . (Rangacarya 1912: 2–3)

A more occupationally focused encomium on the many applications of mathematics appears in another ninth-century text, the *Education of the Secretaries* by the Baghdad polymath Ibn Qutayba. The rapidly expanding 'Abbāsid Empire needed many administrative functionaries skilled in the arts of the scribe. Mathematical knowledge was an important part of these skills, as Ibn Qutayba explained:

[I]t is indispensable for [the secretary] to study geometrical figures for the measurement of land in order that he can recognize a right, an acute, and an obtuse triangle and the heights of triangles, the different sorts of quadrangles, arcs and other circular figures, and perpendicular lines, and in order that he can test his knowledge in practice on the land and not on the [survey-] registers, for theoretical knowledge is nothing like practical experience.

The Persians [i.e., the Sasanians] used to say that he who does not know the following would be deficient in his formation as state secretary. He who does not know the principles of irrigation, opening access-canals to waterways and stopping breaches; [measuring] the varying length of days, the revolution of the sun, the rising-points [on the horizon] of stars, and the phases of the moon and its influence; [assessing] the standards of measure; surveying in terms of triangles, quadrangles, and polygons of various angles, constructing arched stone bridges, other kinds of bridges, sweeps with buckets, and *noria* waterwheels on waterways; the nature of instruments used by artisans and craftsmen; and the details of accounting. (Gutas 1998: 111–12)

In the Islamicate and European descendants of the Hellenistic tradition, these cosmopolitan mathematical professionals—polymaths in every sense of the term—inherited the mantle of classical geometry as well as acquiring and inventing new technical tools. Moreover, their esoteric powers sometimes inspired as much distrust as admiration, especially in the context of astrology. This is exemplified by the legendary career of Michael Scot, university-trained churchman, translator of Arabic astronomical and philosophical works, astrologer, diviner, and dedicatee of a volume of Leonardo (Fibonacci)'s *Book of Calculating*, among other things. A renowned translator and scholar at the imperial court of Frederick II, Scot in the following century cut an ignominious figure among the fortunetellers inhabiting the eighth circle of hell in Dante's *Divine Comedy*, as one who "in life wished to see too far before him . . . Michele Scotto, who mastered every trick of magic fraud."

Indeed, the character of the learned but unsanctified *mathematicus*-magician has had a long and colorful career in European fiction, including the "philosophre" in Chaucer's *Canterbury Tales* of the late fourteenth century, who used his "magyk natureel" to temporarily vanish the great rocks on the coast of Brittany. Referring to his Toledan tables (see the following section), this "subtil clerk" in the Franklin's Tale deduces the astronomically correct moment to create his illusion, for which his client has agreed to pay him a thousand pounds of gold:

What somme sholde this maistres gerdon be	What sum should this master's payment be
To remoeven alle the rokkes of Britayne . . .	To remove all the rocks of Brittany . . .
Lasse than a thousand pound he wolde nat have . . .	Less than a thousand pounds he would not have . . .
His tables Tolletanes forth he brought,	His Toledan tables he brought forth,
Ful wel corrected, ne ther lakked nought,	Accurately corrected, nor lacking anything,
Neither his collect ne his expans yeeris,	Neither the accumulated nor the individual years,
Ne his rootes, ne his othere geeris, As been his centris and his argumentz,	Nor his dates or his other things, Such as his anomalies and arguments,
And his proporcioneles convenientz	And his appropriate proportionals
For his equacions in every thyng . . .	For his equations in everything . . .
Ful subtilly he kalkuled al this . . .	He calculated all this very knowledgeably . . .
And knew ful weel the moones mansioun	And knew very well the lunar mansion
Acordaunt to his operacioun,	Corresponding to his procedure,
And knew also his othere observaunces	And knew also his other observations
For swiche illusiouns and swiche meschaunces	For such illusions and such deceits
As hethen folk useden in thilke dayes.	As heathen folk used in those days.
For which no lenger maked he delayes,	For which he no longer made delay,
But thurgh his magik, for a wyke or tweye,	But through his magic, for a week or two,
It semed that alle the rokkes were aweye.	It seemed that all the rocks were gone.

We can infer from remarks like those of Ibn Qutayba on secretarial training that there existed administrative ideals of generalist mathematical competence which embraced surveying, engineering, accounting, and other computational practices as well as astronomy. Indeed, many individuals officially designated "scribes" or "secretaries" or "accountants" or "astronomers" were learned in more than one of these areas. However, the coordinating and carrying out of complex tasks might also take advantage of specialization in training and employment. In the remainder of this chapter, we examine how certain types of mathematical tasks overlapped in medieval professional practice and how their practitioners were employed and supported. Finally, we discuss the educational systems that practitioners relied upon for mathematical training and credentials.

TIME, THE HEAVENS, AND PREDICTION

The varieties of quantitative celestial prediction might be said to constitute a professional "cluster," both in traditional Chinese astral sciences and in their spherical-universe counterparts employing forms of the Greco-Babylonian zodiac. Some astral science practitioners were responsible for the jobs of calendar-making and timekeeping, which often carried a "research component" in the form of revising and refining the mathematical models describing the cosmos. Some identified ominous heavenly phenomena and foretold their consequences in accordance with elaborate divinatory systems, which might be specialized for medical diagnosis and prognosis. There was seldom a rigid occupational boundary between the mathematical description of celestial phenomena ("astronomy") and the forecasting of fate from them ("astrology"), and many practitioners were experts in both.

Astrology, prognostication, the determination of auspicious and inauspicious days, all have a long history in China and are intimately connected with mathematics. The ancient philosophy known as *wuxing* (five phases) provided an association for each of the five planets with fundamental principles of alchemy and divination as well. Each planet had an association with a color, a fundamental element, and a mythical animal. Venus for example is associated with metal and the white tiger, Mars with fire and a vermilion bird or phoenix, Jupiter with wood and an azure dragon. The five phases were regarded as cosmic elements of change and therefore directly associated with such principles as the *Yijing* and *feng shui*. They were also used to account not only for cosmic cycles but the fate of the empire, changes of political power, the effectiveness of medicines, and interactions of the body's internal organs. In medicine, for example, fire was associated with the heart, water with the kidneys. One's fate could be read in terms of the astrological positions of the planets at one's birth or at crucial moments in both the political history of the empire or the health of a given individual. One major difference between Chinese astrology and Western astrology is that each of the twelve animals associated with the Chinese zodiac, determined by the motion of Jupiter through the heavens nearly once every twelve years, means that each zodiac sign is therefore correlated with an entire year, for example the year of the tiger or monkey, and the signs do not changing regularly throughout the year, roughly correlated with each of the twelve months as in the Western zodiac. But the basic principles of how the stars and positions of the planets influence one's destiny are common to both in determining one's fate, whether of the empire or of the individual.

Fate was also mathematical in that certain numbers were considered to be auspicious—*ba*, meaning eight, for example, is considered lucky because it sounds like another word associated with great wealth and good fortune; likewise *si*, meaning four, is regarded as unlucky because its pronunciation is the same as the character for death. The association between mathematics and divination can be traced back to the *Yijing* (I Ching), the ancient book of divination, the *Book of Changes*, which involved sixty-four hexagrams, each with divinatory significance that could be read directly from the *Yijing* itself. The first step in determining the hexagrams is the manipulation of forty-nine yarrow stalks into various groups—similar to the manipulation of counting rods that were the mainstay of every mathematicians' computational tool kit. Thus the association between the *Yijing* and mathematics was directly due to the resemblance of the manipulations of stalks or rods common to both. As further evidence of this connection, the Tang dynasty philologist Yan Shigu notes that the Chinese symbols for numbers were believed to have been derived directly from the trigrams of which the hexagrams of the *Yijing* are comprised.

The association between mathematics and prognostication is reflected in Chinese almanacs and day books (books of fate), hemerological works devoted to determining "good days" and "bad days." This was as important for the emperor in properly setting the dates for state rituals or insuring the successful administration of the empire as it was for determining the prosperity of the city laborer or bountiful harvests for the country farmer. Such works were used as guides for everyday life—determining the best day for setting an engagement or wedding, when to avoid travelling, making investments, scheduling a funeral, beginning a new business, etc. These are also closely related to the ancient practice of *feng shui*, a form of geomancy used to determine when to move a house or the best directions in which to align a given property. Thus the almanacs did much more than list eclipses, the beginnings of the seasons, or the days when it would be hot or cold, but they also served as ready references to settle important matters of one's day-to-day activities. Given the acknowledged universal significance of numbers as being applicable to virtually everything in the world, it could even be used to predict the sex of an unborn child. The last problem in the *Mathematical Classic of Master Sun* reads as follows: "Now there is a pregnant woman whose age is 29. If the gestation period is 9 months, determine the sex of the child" (Lam and Ang 2004: 223).

The Greco-Islamicate astral sciences based on Ptolemaic models developed and spread with the Islamic empires. Muslim religious observances often depended on the determination of celestial events such as new-moon visibility marking the start of a lunar month, correct times for prayer relative to local sunrise, noon, and sunset, and the local *qibla* or direction of Mecca. Their technical aspects required a good grounding in the geometry and trigonometry of spherical astronomy. Many court astronomers made significant theoretical advances in these fields and in their applications to planetary models; some led substantial state-supported programs or institutions of astronomical observation and research.

By the late thirteenth century, the occupation of *muwaqqit* or timekeeper, demanding expertise in the observational and computational procedures pertaining to Islamic rules on time, the calendar, and the *qibla*, had emerged as a distinct mathematical profession with a prominent role in mosque administration. In the late medieval period these and other mathematical practitioners typically relied upon the rapidly proliferating endowed schools or *madrasas* for their training. Professorships at such institutions largely replaced personal patronage as a source of professional support in mathematical sciences.

Horoscopic astrology, including so-called "mundane" or "political" astrology concerned with political events and the fate of nations, likewise increased in technical complexity and cultural importance. Astrologers were consulted by rulers for routine annual forecasts, on military campaigns, under treatment for illness or injuries, and in many other major undertakings. The casting of a nativity horoscope by a group of court astrologers is a familiar image from Islamic miniature paintings depicting the birth of an heir to the throne (as in Figure 2.2). There is even an account of a woman astrologer, the widow of the 'Abbāsid caliph al-Ma'mūn, who after taking observations with an astrolabe successfully predicted and averted a threat to the life of his successor al-Mu'taṣim. Astrology's professional prestige was somewhat less firm than that of astronomy, however. While many elite scholars endorsed and practiced astrological prediction, at courts or for private patrons, others were skeptical about its validity. And the common street astrologer was evidently somewhat disdained, according to the remarks of the tenth-century mathematician al-Uqlīdisī alluded to in the previous chapter:

> Many a man hates to show the dust board in his hands when he needs to use this art of calculation for fear of misunderstanding from those present who

see it in his hands. It is unbecoming him since it is seen in the hands of the good-for-nothings earning their living by astrology in the streets. (Berggren 1986: 36)

The astral sciences of the Islamicate world provided part of the inspiration as well as sources for the study of these subjects in Byzantium. Remnants of Hellenistic sciences persisted in late first-millennium Greek learning, among them some Greek political horoscopes from late antiquity, the mathematical and astronomical writings of the Athenian Neoplatonist Proclus in the fifth century and the Alexandrian Christian John Philoponus in the sixth, along with the (mostly lost) works of Leo the Mathematician in the ninth. The surviving Hellenistic canon was substantially supplemented in the middle to late Byzantine period by translations and studies of Arabic works, possibly enabled by the influx of Arabic-speaking Eastern Christians after the empire's eastward expansion and recapture of Antioch.

A Greek treatise probably from the eleventh century, titled *On the Mathematical Art* and (pseudonymously) attributed to the famed seventh-century philosopher Stephen of Alexandria, asserts that Christians should study the heavens as not only Muslims but many ancient peoples have done before them. Astronomy and astrology are lauded as key parts of Christian wisdom concerning God's creation. Despite the Orthodox Church's persistent distrust of astrology, subsequent Greek scholars honored these injunctions by incorporating into their scientific corpus a rich mixture of astronomical and astrological material from Hellenistic, Byzantine, and Muslim as well as Jewish and even Latin authors. Their eclectic sources include discussions of the Greco-Islamic astrolabe, commentaries on Ptolemy, translations of Hebrew astronomical tables, excerpts from Hellenistic writers re-translated from Arabic versions back into Greek, and pseudonymous texts ascribed to renowned Greek or Arabic authorities.

By the end of the thirteenth century Byzantine interest in Hellenistic astral sciences had intensified, accompanied in some cases by a sort of "Greek chauvinism" illustrated in the exhortation of the early fourteenth-century statesman and patron of arts and sciences Theodore Metochites: "Be a Greek and shun the theories of the Indians, the Scythians and the Persians or any other foreign ideas." Scholarship in this field was dominated by members of the Orthodox clergy and inspired partly by concerns over the astronomical validity of Christian liturgical calendar, which in the fourteenth century provoked several high-profile disputes on calendar reform.

West of Byzantium in Latin Europe, ecclesiastical institutions were likewise the primary locus for early medieval study of astronomical subjects, in a form rather different from the theoretical spherical astronomy of the late classical quadrivium. The chief aim of such study—in accordance with St. Augustine's

FIGURE 2.1. Miniature showing mathematical instruments employed by Richard of Wallingford, fourteenth-century Abbot of St. Albans, mathematician and inventor of a horologium or astronomical clock. From a manuscript of Thomas Walsingham's *History of the Abbots of St. Albans* (MS British Library Cotton Claudius E IV, fol. 201). Photo by Pictures from History/Universal Images Group / Getty Images.

advocacy of "Christian learning" selectively reorganizing parts of pagan knowledge for doctrinal purposes—was the firm establishment of the Christian calendar and Biblical chronology.

The determination of calendric cycles was called *computus* (Latin "calculation, computation"), especially as it pertained to finding the date of the Easter festival, defined as the first Sunday after the first full moon following the spring equinox. (The problem of appropriately synchronizing luni-solar and weekday cycles was further complicated by differing assumptions about how sacred chronology should count time from Creation.) Technical details of the spherical geometry underlying calendric algorithms were mostly ignored. For the average practitioner entrusted with timekeeping tasks in a monastery or cathedral, *computus* was primarily a matter of ensuring that times of prayer were properly coordinated with changing lengths of day and night throughout the year, and that the liturgical calendar was properly observed.

Starting around the turn of the second millennium, interest in the astrolabe and astrological methods seems to have been the main impetus for increasing assimilation of Greco-Islamic spherical astronomy into Latin learning. The professionals engaged in this endeavor included not only Christian clerics but also traveling scholars and teachers, some of them Jews or converts with detailed knowledge of Arabic sources. Trade and other contacts with Byzantine and Islamic territories (notwithstanding perennial military clashes) facilitated translation efforts in, for example, Antioch, Palermo, Pisa, and Languedoc.

But it was Toledo in the late twelfth century that became the main center of scholarly translation in mathematical sciences in the Latin world. Latin-literate Christians such as the eminent Gerard of Cremona studied Arabic there, working with Arabic-literate Mozarab, Muslim, and Jewish scholars who preserved Toledo's Islamicate heritage of scientific learning. Translations were soon supplemented by new works in Latin and European vernaculars, chief among them the astronomical tables sponsored by King Alfonso X of Castile in the thirteenth century to revise and update the eleventh-century "Toledan Tables."

These "Alfonsine Tables" remained at the heart of European astronomical practice until displaced by Copernican heliocentrism. Various revisions and reworkings of their content testify to the interest in advanced study of astronomical and calendric models manifested in Western Europe during this time, despite the universities' mid-fourteenth-century contraction due to the depredations of the plague. Such study could now constitute a professional career in its own right, underpinned by a variety of occupational roles, as illustrated by the activities of the astronomer and music theorist John of Murs in the first half of the fourteenth century. After attaining the degree of master of arts at the University of Paris, John lived and worked at various times as a university scholar, a church canon, and a resident at the courts of the king of Navarre and Pope Clement VI in Avignon.

Astrology, including its medical applications, was an important component of astronomical learning, taught sometimes in university courses on astronomy and sometimes in ones on medicine. Earlier Christian abhorrence of this art had been partly overawed by the impressive array of technical mathematical detail that astrological practice could boast, although it was still condemned as illicit magic by many theologians. Philosophers countered with appeals to the subject's scientific basis in predictable celestial phenomena, as opposed to mystical incantations and spells. The general respectability and prestige of the court physician-astrologer are reflected in the character of Chaucer's "Doctour," who

FIGURE 2.2. Miniature of a group of Muslim and Hindu astrologers (below) casting a nativity horoscope for the newborn Mughal prince Salīm, later Emperor Jahāngīr (above). The bearded Muslim scholar at right holds a Persian document while the Hindu one to his left is drawing an Indian birth chart. A ring dial and a sinking-bowl water-clock provide astronomical data, while the court lady at left is presumably announcing the delivery (Sarma 2008: 84–5). From an early seventeenth-century Mughal miniature painting (Museum of Fine Arts, Boston, "Birth of Prince Salim"). Courtesy of the Boston Museum of Fine Arts / Wikimedia Commons.

was qualified "To speke of phisik and of surgerye, For he was grounded in astronomye."

A learned Latin astronomer familiar with numerous works translated from Arabic, who might even read Arabic himself, would probably have a much more remote acquaintance with the astronomy of the fabled "wise men of India." For example, the legend quoted in Chapter 1 from Abraham ibn Ezra about the supposed visit of the Indian scholar "Kanaka" to Baghdad to teach the decimal place-value number system continues:

> Then, from this scholar, with the Jew as Arabic-Indian interpreter, a scholar named Jacob ben Sharah translated a book containing the tables of the seven planets, all the procedures for the earth, the rising times of the zodiacal signs, the declination and the ascending degree of the ecliptic, the arrangement of the astrological houses, knowledge of the upper stars, and eclipses of the luminaries. But there is no mention in his book of the reasoning in all these matters: only facts are given as matters of tradition. (Goldstein 1996: 12)

Indeed, this description conveys a reasonably accurate general impression of what an Indian practitioner of astronomy and astrology (*jyotiṣa*) would need to know for his (or perhaps occasionally her) work. Traditional Indian calendrics was a highly involved subject, tracking five simultaneous time-unit cycles (the weekday, the lunar-phase day or *tithi*, the lunar mansion or *nakṣatra*, and two other measures of luni-solar motion) in an annual calendar called the *pañcāṅga* ("five-part").

Certain moments in all these cycles, as well as celestial events such as eclipses and other specific configurations of the sun, moon, and planets, had ominous significance. The Indian *jyotiṣa* practitioner had to be familiar with a complex system of predictive astrology including not just nativity horoscopes, but also catarchic astrology (finding auspicious times for performing desired acts, with special emphasis on marriage and military expeditions) and interrogations (prediction of desired information based on astrological analysis of the moment when he question was asked). Various Indian techniques for these subjects were adopted in Islamic, Byzantine, and Latin astrology, while Indian astrologers in their turn expanded their repertoire with some Arabic-Persian developments.

Every village, court, and private patron needed an annual *pañcāṅga* to schedule ritual practices prescribed for astronomically determined times, and to avoid inauspicious or prohibited times for other activities. Randomly occurring mundane events such as births of course also required astrological analysis. A *jyotiṣa* professional of *brāhmaṇa* lineage might be supported, for instance, as a court astronomer-astrologer, as a member of a priestly or monastic Hindu or

FIGURE 2.3. The twelve signs of the zodiac labeled with Hebrew names surrounding the chariot of the sun, from a sixth-century mosaic pavement in the Bet Alpha synagogue in the Bet She'an Valley in Israel. The solstices and equinoxes depicted in the four corners are also labeled with their corresponding month names in the Hebrew calendar. Photo by DeAgostini / Getty Images Stock Photo.

Jain community receiving tax revenues or donations, or by fees earned from casting horoscopes or performing rituals or teaching students.

Some inscriptions record significant rewards from patrons for individual feats of astronomical expertise. One copper-plate grant from central India dated 1128 testifies to the impression made on the ruler Ratnadeva II by a court astronomer's successful eclipse prediction:

> [T]he illustrious Padmanābha . . . has crossed the ocean of astrology . . . declaring in the assembly of the illustrious Ratnadeva, in the presence of all astronomers, that [on a specified day] during the third quarter of the night

> with [the moon in] the constellation Rohiṇī, there would be a complete eclipse of the moon . . . [T]he king, who was pleased, gave him as a grant the village of Ciñcātalāī in the area of Anarghavallī with all taxes. (Montelle 2011: 183)

The spectacular aspect of such feats was also valuable for professional prestige. The astronomer-mathematician Brahmagupta described in his treatise the procedure of setting up a sighting-line with two poles of different heights to guide the viewer's eye to the exact celestial position where the astronomer had predicted "an eclipse or a conjunction with a planet, for the purpose of astonishing the people or the king."

Every Chinese emperor was expected to "grant the seasons" to his subjects by means of annual almanacs. The mathematical-astronomy systems underlying them were a means of establishing and maintaining imperial legitimacy, especially by improving upon the accuracy of earlier systems; in consequence, some seventy systems were used between the start of the Common Era and the modern period. The most prominent of all the Chinese dynastic calendars, the Season-Granting Calendar, was adopted in 1280 at the beginning of the Mongol Yuan dynasty. As the culmination of the long history of Chinese calendars, it was regarded so highly that it was adopted with only minor changes as the *Datong li* (Great Concordance Calendar) by the Ming dynasty in the fourteenth century. Even after the arrival of the Jesuits in China and the prominent role they began to play in the Astronomical Bureau (including the 1653 imperial proclamation of Adam Schall von Bell as the leader of the Bureau), the Season-Granting Calendar remained the most highly regarded of the traditional calendars. Its fame spread beyond China: it was officially adopted in Korea in 1309 and was well known also in Japan, where it was studied by the distinguished eighteenth-century mathematician Takebe Katahiro.

Through the centuries, the work of "granting the seasons" required numerous specialists throughout the Chinese state bureaucracy. As early as the Eastern Zhou dynasty and as late as the Qing, records refer to students of the *chouren*, hereditary officials responsible for matters dealing with astronomy, the calendar, and the construction of almanacs. At the top of the list stood the Grand Clerks, followed by retinues of specialist observers, record-keepers, calculators, and omen-interpreters, all working together.

COMMERCE, CONTRACTS, AND ADMINISTRATION

As medieval trade and bureaucracy grew, and with them the size of individual personal fortunes, so, too, did the mathematical tools they employed.

Commercial and administrative computation developed as a professional specialization beyond basic scribal literacy.

Accounts of administrative roles among government officials in China reveal, for instance, that "accounting managers" were responsible for keeping the accounts of the civil service, including records of government income and expenditures, taxes due on property, and the compilation of various statistics related to affairs of state. Military matters were the province of special "accountant officials" within the army, who kept track of military expenses, weapons, rations, and payments to the troops.

Textbooks such as the one written by the Indian mathematician Śrīpati in the eleventh century indicate the complexities of trade and commerce, and the foundational mathematical knowledge practitioners needed to manage loans, trade, barter, and compute wages and prices. Śrīpati prescribed rules for performing various financial transactions, which he illustrates by (unsolved) examples shedding some light on their professional context. One such example, encouragingly addressed in the second person to the student, describes a loan problem with some of its overhead costs:

> Clever man, the interest is five [*drammas*] on one hundred per month. [The fee] for the guarantor (*bhāvyaka*) is one [*dramma*], one half [of a *dramma*] for the calculator (*vṛtti*), and similarly, one fourth [of a *dramma*] for the scribe (*lekhaka*) [all per month]. Friend, [the total amount] is nine hundred and five in twelve months. State the capital and so on if you are clever at determining mixtures. (Dhammaloka 2019: 260)

Besides the lender, borrower, and loan guarantor, this transaction requires a scribe to record the details and a "calculator," presumably an accountant making sure the correct interest is applied and the various charges are computed and administered properly. Śrīpati's statement that the guarantor's commission is twice that of the calculator, which in turn is twice that of the scribe, indicates that in some contexts numeracy has a higher market value than literacy (although possessing capital still pays best of all).

Along with trade and government administration of the sort outlined in Ibn Qutayba's previously quoted description of secretarial knowledge, many mathematical professionals in medieval Islamicate societies paid significant attention to the legal sphere of inheritance and legacy distribution. The complex Islamic laws of inheritance required expertise in computing the distribution of legacies according to multiple legal stipulations. The ninth-century *Algebra* of al-Khwārizmī was described by its author as addressing the requirements of inheritance calculation, as well as various branches of engineering:

> [The caliph] al-Ma'mūn . . . has encouraged me to compose a short work on "Calculating by (the rules of) Completion and Reduction" (*al-jabr wa'l-muqābala*), confining it to what is easiest and most useful in arithmetic, such as men constantly require in cases of inheritance, legacies, partition, law-suits, and trade, and in all their dealings with one another, or where the measuring of lands, the digging of canals, geometrical computation, and other objects of various sorts and kinds are concerned. (Rosen 1831: 3)

The following scenario in a sample problem from the *Algebra* illustrates some of the complexities of the necessary knowledge, both legal and mathematical:

> Suppose that a man, in his illness, emancipates two slaves, the price of one of them being three hundred *dirhams*, and that of the other five hundred *dirhams*; the one for three hundred *dirhams* dies, leaving a daughter; then the master dies, leaving a daughter likewise; and the slave leaves property to the amount of four hundred *dirhams*. With how much must everyone ransom himself? (Rosen 1831: 146)

Many of the monastic foundations and wealthy estates of Europe functioned as centers of travel, trade, and finance, with large and diverse economies of their own which required some form of quantitative account-keeping. Institutional account books might contain lists of receipts and expenditures, tables of prices and wages, asset inventories, and periodically calculated balances. As travel and trade connections with Islamicate neighbors increased, techniques for managing credit, insurance, debt, endowments, and other financial concepts were transmitted and modified. In the Latin world, treasurers and moneychangers were joined by bankers, insurers, investment managers, and financiers in general. (Although Muslims, Christians and Jews all acknowledged some form of doctrinal prohibition of usury in certain contexts, such prohibitions could be more flexible in practice, especially for transactions between members of different faiths.) Increasingly detailed and formalized accounting practices gave rise in Italy around the early thirteenth century to the double-entry bookkeeping methods that became standard in banking and commerce thereafter. By the fourteenth century, the cashless bill of exchange had developed into the primary European instrument for carrying out international payments, with a sideline as an instrument of credit allowing interest charges to dodge Church laws against usury.

SURVEYING, ART AND ARCHITECTURE, ENGINEERING

Descriptions of how medieval practitioners applied mathematics directly to the physical world frequently focus on building (including civil engineering), surveying, and instrument construction. The characterization of these activities as specifically "mathematical professions," rather than simply a few of the many occupations requiring some degree of everyday numeracy, is largely due to the prominence of geometry in their subject matter. Yet their practices were often considered quite distinct from academic geometry. As the sixth-century Roman statesman Cassiodorus remarked concerning land surveyors (*agrimensores*):

> The professors of this science are honored by a more earnest attention than falls to the lot of any philosophers. Arithmetic, Theoretical Geometry, Astronomy, and Music are discoursed upon to listless audiences, sometimes to empty benches. But the land surveyor is like a judge, the deserted fields become his forum, crowded with eager spectators. He shows what he is saying. He proves what he has learned . . . (Merdinger 1954: 100)

The *agrimensores*' role dealt with interpreting land maps and plans, determining legal boundaries of properties, computing their areas, and resolving land disputes. Their surveying techniques and tools were based on those of late antiquity, explained in Roman manuals of "gromatics" (so called from the traditional *groma* or surveyor's pole). There is no evidence of the use of trigonometry or other advanced geometry methods in medieval surveying: practitioners relied on proportional sides of similar triangles to determine heights and distances. Indeed, angles other than right angles are rarely mentioned in surviving texts.

The late medieval practical-geometry manuals mentioned in the Introduction introduced refinements such as sighting with the astrolabe and the double-difference method, although it is not clear when and to what extent they displaced traditional methods in actual practice. Some surviving Islamic texts likewise explain how a surveyor could use astrolabe sightings to calculate heights, depths, and distances, or how to use leveling instruments in hydraulic surveys for irrigation works. The prescriptions of Ibn Qutayba suggest that elementary geometry sufficed for "surveying in terms of triangles, quadrangles, and polygons of various angles."

Surveying instruments such as the compass, carpenter's square, leveller, plumb-line, measuring poles, and even pedometers had been used in China since ancient times. But after the third-century *Sea Island Mathematical*

Manual of Liu Hui explaining the double-difference technique and an elegant three-pole variation described in the sixth century by Zhen Luan for measuring the width of a river, mathematical comprehension of surveying geometry lapsed somewhat until revived in the Song period. Again, practical surveyors had their own toolkit of methods that were applied in the field, relying more on idealized simple rectangular grids than on more precise geometric modeling.

In the Song dynasty, problems of surveying or distance determination were often used as a pretext to demonstrate more advanced mathematical techniques, as was the case in the classic *Sea Mirror of Circle Measurements* published by Li Ye (also known as Li Zhi) in 1248. There, 170 problems are presented, usually in terms of a person walking from one point to another along roads at right angles to each other, usually tangent to or through the walls of a circular fort or town. All of the problems are the same, namely to find the diameter of the circle, given two of the three sides of any given right triangle a, b, c, where c represents the hypotenuse. And in every case the answer is the same: 240 *bu*. Thus the given conditions come down to two equations, $a^2+b^2=c^2$ for the given right triangles, and $a+b=c+d$ for the triangles and diameter of the circle, which itself is inscribed in both a right triangle and a square. A typical problem takes the form: if A walks 135 *bu* south from the south gate and B walks 16 *bu* east from the east gate, they can just see each other. What is the diameter of the circle? All of the problems in the *Sea Mirror of Circle Measurements* are solved using the *tianyuan* or "heavenly element method." Notably, Li Ye's book is the oldest extant work in which this method is explained.

In the assorted subfields of construction and engineering, too, it is not always easy to identify the overlaps and divergences between mathematical training and practical craft. On the one hand, it is easy to perceive the abstract geometric principles embodied in some artisan tools, such as the medieval builder's "cord of thirteen knots," with its twelve equally spaced intervals that can be efficiently transformed into a 3–4–5 right triangle and other figures. Moreover, academic mathematicians might be called upon for expert advice in building projects, like the sixth-century Byzantine geometers Anthemius of Tralles and Isidore of Miletus, who were consulted in the rebuilding of Constantinople's Hagia Sophia cathedral and the creation of flood control structures at Dara on the Persian border.

Statics or "the study of the balance" was a recognized field in Islamicate practical mathematics in the second millennium, and professionals serving in such a capacity were given the title "al-Qabbānī" or "the one who works with and builds balances." Problems in transport of loads, counterbalancing, centers of gravity, and the pulling of heavy bodies were analyzed in a new class of texts pitched to these professionals, such as the

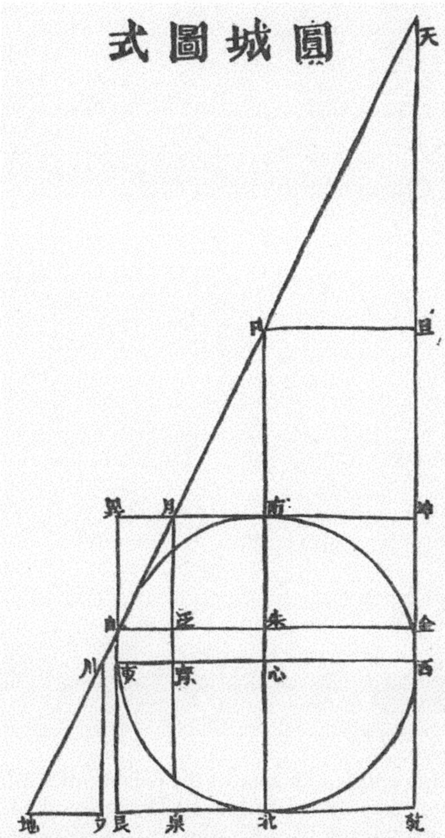

FIGURE 2.4. Diagram of a circular city, with Chinese characters, some of them directional (such as North, South, East, West), denoting critical points on the diagram. Other characters are used from the *Yijing* (*Book of Changes*). All of the 170 problems in this work are cast in terms of different parts of this diagram. From a 1798 edition of the *Sea Mirror of Circle Measurements* by the Song dynasty mathematician Li Ye (Li Zhi), first published in 1248 (Needham 1959: 44). Courtesy of the Needham Research Institute, Cambridge, UK.

eleventh-century *Guiding the Learned Men in the Art of the Steelyard* by the Iranian al-Isfizārī.

On the other hand, the exponents of geometric theory and practical craft sometimes talked past each other, due to fundamentally different perceptions of the nature of their task. A rare contemporary report of such an encounter was recorded by the Baghdad astronomer-mathematician Abū al-Wafā' al-Būzjānī in the tenth century, noting the different approaches taken to geometric constructions for elaborate architectural tiling patterns by geometers and by the craftsmen who actually made the tilings:

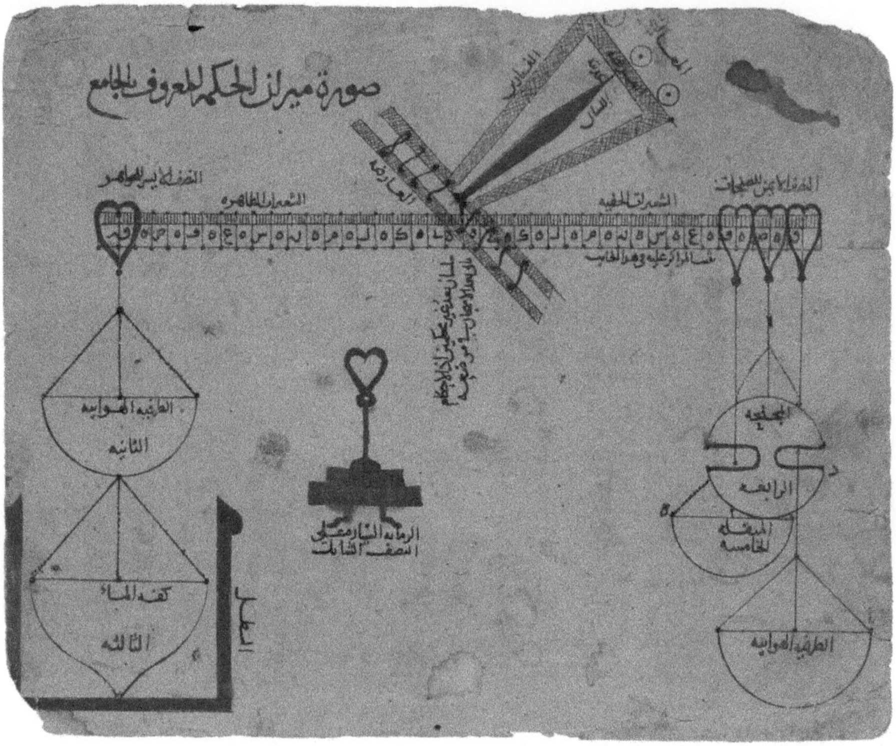

FIGURE 2.5. Illustration entitled "Figure of the balance of wisdom" of an elaborate weighing device developed by Abū al-Fatḥ al-Khāzinī around 1120. From a thirteenth- or fourteenth-century manuscript of al-Khāzinī's Arabic *Book of the Balance of Wisdom* (Lawrence J. Schoenberg Collection, University of Pennsylvania, LJS 386, p. 192). Image courtesy of the University of Pennsylvania.

> A number of geometers and artisans made errors in the matter of these squares and their assembling. The geometers made errors because they don't have practice in applied constructing, and the artisans because they lacked knowledge of reasoning and proof . . .
>
> I was present at a meeting in which a number of geometers and artisans participated. They were asked about the construction of a square from three [identical] squares. Geometers easily constructed a line such that the square of it is equal [in area] to the three squares, but none of the artisans was satisfied. They wanted to divide those squares into pieces from which one square can be assembled. (Sarhangi and Jablan 2006: 10)

Mostly, however, what we know about the interaction of academic geometry with artisanal craft, including instrument-making, comes from geometric problems and "geometrized" versions of craft rules preserved in academic textbooks. Indian mathematical treatises, for instance, generally devoted a section to the various *vyavahāras* or "practices" of computation, including excavation (*khāta*), stacking (*citi*), sawing (*krakacika*), and heaps (*rāśi*). The formulas prescribed for these practices often incorporated multiple versions to take into account the characteristics of different materials, as in the following example distinguishing between the units appropriate to excavating stone rather than earth:

> [In the case of stone excavation,] when the area is multiplied by the thickness (*piṇḍa*) as previously stated, [the result] is the cubic cubits of stones. These, multiplied by nine [and] divided by four, are accordingly the stone-cubits. (Dhammaloka 2019: 109)

EDUCATION AND TRAINING

Given the diversity of medieval mathematical practice and profession, how did students prepare themselves to work in these fields? Craft practitioners, including makers of mathematical instruments, worked in some capacity under an accomplished master for hands-on training and access to the necessary materials. In this section, we discuss instead some of the features of what might be called general academic mathematics education in different cultures and periods, and the processes employed for specialization and credentialing.

Educational Institutions

The institutions that supported and perpetuated mathematics in China were largely associated with the imperial administration, requiring specialists in tasks ranging from the maintenance of tax records to the astronomical labors upon which the creation of dynastic calendars and seasonal almanacs depended. Other bodies were devoted to the production of teaching manuals, books for specialists, or the creation of tables used by astronomers.

Government schools in which mathematics at least intermittently formed part of the established curriculum date back to the sixth-century Sui dynasty. According to one official account, the Imperial Academy in the Sui era had two mathematics teachers and two assistants for eighty students. The Sui also revived the empire-wide civil service examination system, which from time to time included test material assessing students' mastery of canonical mathematical works.

Schools in the Tang dynasty recruited students from among the population at large, not just bureaucratic elites, for a seven-year course of study marked by yearly written and oral tests. The parallel tradition of mathematics teaching in private academies went back at least to the academy of Hu Yuan in the eleventh century. Mathematics teaching is attested in, among others, the Ziyang Academy at Suzhou, the Longcheng Academy at Changzhou, and the Fenglong Academy in Hebei province. In the thirteenth century one renowned Fenglong instructor, Li Ye, drew thousands of disciples to his institution. The emperor Kublai Khan held Li Ye's counsel in such high esteem that he made several attempts to bring Li Ye to the imperial capital Beijing, and succeeded in doing so briefly in 1264 with an appointment at the Hanlin Academy. But Li Ye did not find the Hanlin Academy nor the Beijing capital congenial, and after only a few months returned to the Fenglong Academy, where he died in 1279.

An independent itinerant lecturer might instruct students who wanted to learn mathematics from a "master." The twelfth-century mathematician Qin Jiushao studied with one such "recluse scholar," and a century later Zhu Shijie taught pupils in this capacity. Around the middle of the second millennium, the manipulation of the bead abacus or *suanpan* was taught by professional "experts" unaffiliated with an institution. Finally, there were self-professed autodidacts unable to find a suitable master and relying on what books they could find to teach themselves, such as Gu Yingxiang in the sixteenth century and even some nineteenth-century Qing-era scholars.

Private instruction, especially within *brāhmaṇa* extended families with hereditary occupations in the mathematical sciences, was the main framework for mathematical training in the Sanskrit ecumene. Beginning at an age anywhere from shortly after five years to early teens, depending upon his hereditary social category and educational aims, the pupil studied a standard program of Sanskrit learning for about twelve years at the home of the teacher, who might be his father or uncle or other relative. Small-group cohorts of residential pupils were often taught together, and the course of study might be prolonged for additional years depending on the student's needs or interests.

Rulers, temples, patrons, and Buddhist and Jain monasteries also supported some more formally organized schooling. Probably the most famous such endowment as far as mathematics teaching is concerned (although almost no knowledge survives about its operation or duration) was the educational foundation or *maṭha* described in a thirteenth-century inscription as established by a grandson of Bhāskara II, to promote the study of his illustrious ancestor's works along with those of other family members. This school was funded by a grant of land from the local ruling family, at whose court Bhāskara as well as his son and grandson had served, in commemoration of a lunar eclipse (presumably one that had been successfully predicted to nullify its inauspicious effects). The inscription also acknowledges donations of land and other resources from other patrons, and describes in a local vernacular the officially decreed tax levy to support the school:

> Out of every hundred areca nuts purchased, the purchaser was to donate five to the *maṭha*. Likewise a part of oil seeds brought to the oil mill would go to the *maṭha*, as also one ladle of oil from each oil mill should go to the *maṭha*, and so on. (Hayashi, Montelle, and Ramasubramanian 2019: 14)

Chinese Buddhist visitors to India around the middle of the first millennium recorded some descriptions of large monastic foundations, especially at Nalanda in modern Bihar, supporting and teaching thousands of advanced students in the study of Buddhist works together with aspects of *brāhmaṇa* learning including astronomy. Jains also had monasteries and *maṭhas*, sometimes patronized by Hindu rulers, where secular subjects including Sanskrit mathematics and astronomy might be taught alongside Jain doctrine and philosophy.

Mathematics teaching in the Islamicate world up to the twelfth century likewise was not primarily "institutional." Private tutors taught at their students' residences or held small classes in their own homes. The general education mentioned in Chapter 1 might occupy pupils for several years beginning at the age of six; age fourteen or thereabouts marked the beginning of more advanced specialist schooling, which might continue well into adulthood. Scholars of established reputation were more likely to be employed by wealthy or even royal families, while those in more modest circumstances hired lesser-known itinerant teachers, whose proficiency might be less satisfactory.

Shortly after the turn of the second millennium, this education began to shift from private residences to more public religious spaces such as schools and mosques. Endowed school buildings or *madrasas* were founded beginning in the eleventh century, and within a few decades had started to incorporate teaching in the mathematical sciences as well as religious instruction. This system, which provided financial support for both teachers and students, still relied chiefly on individual instruction; (adult) students might travel long distances to a chosen *madrasa* based on the presence of a particular teacher there.

Syriac, Coptic, Greek, and other non-Muslim students in the Islamicate world might be taught with Muslim students by a tutor, or might study at a monastery or other religious institution. Byzantine Greeks also typically learned mathematical sciences, if at all, from a private instructor. Meanwhile, Western Europe's monastic and cathedral schools gave rise to communities of teachers and learners known as scholastic guilds. These in turn became incorporated as the early universities from the eleventh century onward, beginning with foundations at Bologna, Paris, and Oxford.

The role of monasteries in mathematical pedagogy was diminished with the advent of universities in the late medieval period, but monastic scholarly activity continued and even established a sort of symbiosis with its university counterpart.

For instance, a papal bull of 1336 required all Benedictine monasteries to send a 5-percent quota of their monks to university. Autodidactic study was still practiced at monasteries by many inquiring scholars, like the non-university-trained John Westwyck of St. Albans in fourteenth-century England.

Less academic forms of instruction emphasizing commercial and administrative mathematics were available from itinerant teachers. On-the-job training in the increasingly international mercantile realm was provided via the duties and travel associated with family businesses like that of Leonardo of Pisa (Fibonacci). By the late thirteenth century, commercial mathematics education in Italy began to be institutionalized as the "abbaco" or "abacus" school, teaching practical computation centered on decimal place-value arithmetic and accounting problems. The instructional medium was vernacular rather than the Latin of the universities, and the pupils were usually around eleven to thirteen years old, including budding architects, engineers, and artists as well as merchants.

FIGURE 2.6. Financial transactions in the Banco di San Giorgio in Genoa, illustrating "Avarice" in a fourteenth-century Latin treatise on the seven capital vices, written in Genoa's Cocharelli family (MS British Library Additional 27695, fol. 8r). Courtesy of the British Library, London, UK / Alamy Stock Photo.

Curricula, Textbooks, and Credentials

In all medieval mathematical traditions, the main instrument of learning was the authoritative treatise, buttressed by the lectures and commentary of the teacher. Standard textbooks in the Islamic world included Arabic translations of Euclid's *Elements*, Ptolemy's *Almagest*, and Aristotle, along with the so-called "Middle Books" excerpts on subjects like spherics, optics, and aspects of astronomy. These readings might be supplemented by a variety of other excerpts from classical texts like the *Introduction to Arithmetic* of Nicomachus, philosophical and astrological works, or specialized didactic material composed by the teacher himself or a contemporary author. Advanced students progressed to higher geometry and astronomy, typically including the *Conics* of Apollonius. In addition to their systematic exposition of assigned texts, teachers would meet with small groups of pupils to work through and discuss problems. And it was the individual teacher rather than the *madrasa* or other institution who officially awarded the credential that students earned at the conclusion of study, a certificate called an *ijāza* qualifying them to teach what they had learned.

Their contemporaries in Latin monastic schools and universities pored over the standard textbooks of the trivium and quadrivium, including early medieval works such as *The Marriage of Philology* and *Mercury* by Martianus Felix Capella, *On the Liberal Arts* by Aurelius Cassiodorus, Boethius' *On Arithmetic*, and passages from Isidore of Seville's compendious *Etymologies*. Beginning around the end of the first millennium, students might also play a board game called *rithmomachia* strengthening their knowledge of figurate numbers and different kinds of ratios. After about six years of study with a number of different specialist teachers, during which the student would have earned his *artium baccalaureus* (BA) status, the university conferred upon him the degree of *artium magister* (MA), qualifying him to teach the liberal arts or to proceed to advanced study in law, medicine, or theology.

From about the twelfth century, Latin translations of works in the Arabic mathematical curriculum were employed in university teaching beyond the classical quadrivium, e.g., in advanced classes in astronomy, astrology, or medicine. New subjects introduced in this period included the exotic "algebra and almuchabala" and the "study of triangles," a field later known as "trigonometry." Beginning at the University of Paris, curricular offerings were further expanded in the fields of calendrics, time-reckoning, and astrology.

These relied on three key new textbooks by a thirteenth-century English cleric on the Paris faculty, John of Holywood (or Sacrobosco): *Algorismus* on decimal place-value arithmetic, *On the Sphere of the World*, an introduction to spherical astronomy and cosmology, and *On the Reckoning of the Year, or Church Computus* concerning calendar computation. As this educational trend spread among European universities, "research publications" such the Alfonsine Tables were incorporated into the enlarged curriculum.

The growing willingness in the late medieval Greek and Latin worlds to reclaim "pagan" science of classical antiquity, further enriched by other non-Christian traditions in the meantime, is illustrated by an intriguing academic contest that took place during an embassy of scholars from King Conrad IV of Germany to the Emperor of Nicaea in Byzantium in 1253. The Emperor's son, Prince Theodoros, described in his correspondence a debate held in Greek between the Western visitors and their Byzantine counterparts to compare their proficiency in science. A problem in spherical geometry that had been submitted by Theodoros' teacher Nikephoros Blemmydes was pronounced by all the contestants to be impossible, but thanks to Blemmydes' instruction, Theodoros himself was able to solve it. The prince commented unfavorably on the pride of the Latin delegates in their learning on "geometry and astronomy, arithmetic and music" and numerous other subjects, which their achievements apparently did not justify.

The abacus schools of the period of course had a much more commercially focused curriculum requiring no knowledge of Latin, much less Greek. Pupils studied manuals in Italian which explained the essentials of finger-reckoning, decimal place-value arithmetic of integers and fractions, basic geometry and the Rule of Three, the Florentine currency system, algebra techniques, and sample problems testing the students' application of their learning.

There was no such institutionalized curricular division between "commercial" and "academic" subjects in the Sanskrit textbooks of Indian mathematics, though individual students doubtless prepared for different occupational specializations. Standard textbooks like the famous *Līlāvatī* of Bhāskara II covered subjects ranging from basic arithmetic operations to commercial mathematics and the practical geometry of the *vyavahāras*, along with more advanced topics in geometry, series, and combinatorics. A student who had mastered arithmetic would go on to algebra, and those intending to practice astronomy or astrology would study a treatise on *jyotiṣa*. Oral instruction centered on verse treatises remained the standard teaching format at all levels, and tradition required pupils to memorize Sanskrit verse formulas in order to refer to them more readily.

A *jyotiṣa* treatise often presented its computational formulas (*gaṇita*) in its first part and followed with a second part on the theory and geometry of spherical astronomy (*gola*, "sphere"). Today's mathematics teachers can empathize with some authors' pleas that students should not neglect the deeper understanding to be obtained from the latter section once they had memorized the algorithms of the former! Such admonitions were often motivated by warnings that a student would need theoretical understanding to avoid humiliating displays of ignorance in "assemblies of the learned." Mathematical texts do not supply many details about this custom of "assemblies," which was also practiced within other *śāstras* and seems somewhat similar to the modern

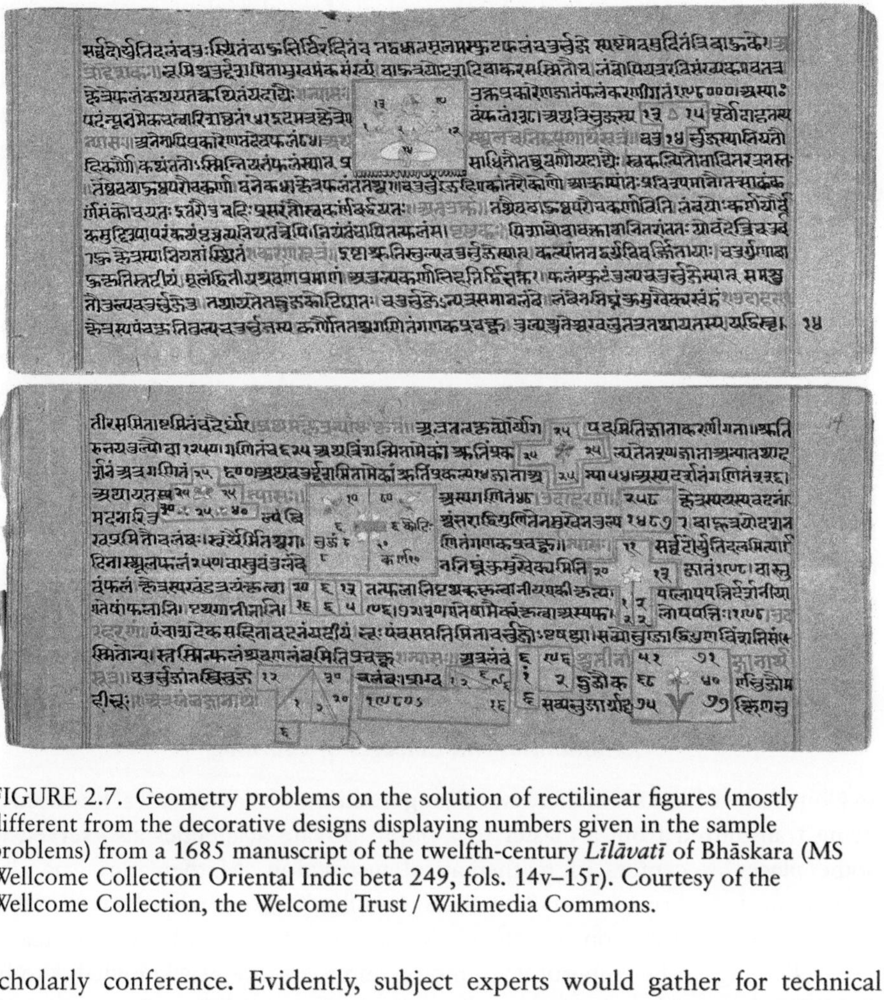

FIGURE 2.7. Geometry problems on the solution of rectilinear figures (mostly different from the decorative designs displaying numbers given in the sample problems) from a 1685 manuscript of the twelfth-century *Līlāvatī* of Bhāskara (MS Wellcome Collection Oriental Indic beta 249, fols. 14v–15r). Courtesy of the Wellcome Collection, the Welcome Trust / Wikimedia Commons.

scholarly conference. Evidently, subject experts would gather for technical discussion of problems and discoveries, sometimes supported by court sponsorship, and a promising young scholar would have the opportunity to impress his elder colleagues and contemporaries (and perhaps also potential patrons?) with his mastery of the discipline. No more formal accreditation procedure for fledgling mathematical practitioners is known to have existed, other than the Sanskrit pupil's usual prescribed ceremony for concluding the period of residential study, including a gift presented to the teacher.

The standard medieval Chinese mathematical curriculum known as *The Ten Mathematical Classics*, comprising the *Mathematical Classic of the Zhou Gnomon*, the *Nine Chapters*, the *Sea Island Mathematical Manual*, and several other classic and contemporary works, was presented to the Tang Emperor in 656. According to a later history of that era, its compilation stemmed from bureaucratic reform efforts at the beginning of the dynasty:

> A lot of the arguments in the *Mathematical Manual of the Five Government Departments* and others among ten [authoritative] mathematical manuals are contradictory. Li Chunfeng together with the Imperial Academy scholar in mathematics Liang Shu, the teaching assistant Wang Zhenru, and others were ordered to write commentaries on ten mathematical manuals . . . After the completion of these books Emperor Gaozu decreed their use throughout the State Institutes. (Li and Du 1987: 105)

The curriculum took seven years to complete, with students divided into two tracks. The less advanced group spent three years on the *Nine Chapters* and the *Sea Island Mathematical Manual* alone. The subject of mathematics was eliminated from the curriculum of the Imperial Academy altogether in 658, and passed into the Astronomical Bureau. In 662 it was again reestablished briefly with one instructor in the Academy, only to be transferred to the Institute of Records a year later. *The Ten Mathematical Classics* were re-edited and printed for the first time during the Northern Song dynasty in 1084. Nonetheless, after 1113, mathematics again disappeared as a subject in the imperial examination system due to complaints about its lack of bureaucratic usefulness, and did not return to government schools until the late nineteenth century under the Qing.

Communities and Lineages of Learning

In China, as we have seen, there was a long tradition of mathematical knowledge being transmitted from a master to a pupil; private academies also served in some places as centers of mathematical instruction. Mathematics was also communicated through "mathematical families," those who had come to specialize in computations of calendars that were considered skills critical to the well-being of the emperor and empire, and therefore was knowledge to be kept secret, passed down from father to son. This tradition of mathematics passed down from father to son was most often connected with the Astronomical Bureau. The earliest known example of such a lineage is that of the fifth- and sixth-century calendrical specialists Zu Chongzhi, his son Zu Huan, and his grandson Zu Hao. The early Qing mathematical clan comprising the brothers Mei Wending, Mei Wennai, Mei Wenmi, and various sons and grandsons, also contained accomplished astronomers.

Renowned scholars also attracted pupils from outside the family, sometimes far outside: it is said that the student Liu Xiangkui sold all his possessions and travelled a thousand *li* to study with Mei Wending. The Northern Song dynasty Confucian philosopher and teacher Hu Yuan revived the teacher–disciple relationship of ancient times. Teaching first in the Suzhou region, and later at the Imperial Academy in the Northern Song capital of Kaifeng, he had thousands of student followers in the course of his lifetime, and among other subjects taught mathematics in his academy.

Some of the notable mathematical lineages of India have been mentioned in the Introduction and earlier in the present chapter; the hereditary occupational groups within which most scholars practiced made the "mathematical clan" a very usual phenomenon. It may be that Bhāskara II's mathematical kin mentioned earlier in this section should be expanded to include the eponymous addressee of his arithmetic text *Līlāvatī*, whom legend holds to have been his daughter. Although it was rare for mathematical works to be addressed to (or, presumably, studied by) girls or women, there are some other female vocatives attested in some such texts. Certainly Bhāskara's tone seems paternally affectionate in his verses apostrophizing a "dear woman," or "dear girl," or "charming woman." Other verses, however, are addressed in masculine vocatives to a "skillful calculator," "friend," "mathematician," or "merchant," suggesting that Bhāskara's treatise may in fact be a compilation of materials composed for a variety of students.

This sort of mathematical lineage is less common but hardly unknown in the Islamic world. Besides the three ninth-century Persian "sons of Moses" or Banū Mūsā brothers who collaborated on works discussing geometry and mechanical devices, there was, for example, the tenth-century Sabian geometer Ibrāhīm ibn Sinān ibn Thābit ibn Qurra of Baghdad, whose physician-astronomer father Sinān ibn Thābit was the son of the illustrious translator and mathematician Thābit ibn Qurra.

Again, wives and daughters are far less common in such Islamicate mathematician kinships than brothers or sons, but not totally absent. The twelfth-century historian Ibn Bashkuwāl left profiles of some 1500 prominent male scholars and 17 female ones, most of whom were connected with one of the learned men by blood or marriage. Of one of them, the tenth-century Lubna of Córdoba, Ibn Bashkuwāl recounts that she was the palace secretary of the Umayyad caliph al-Ḥakam ibn 'Abd al-Rahmān and an excellent writer and poet; "her knowledge of mathematics was also immense and she was proficient in other sciences as well."

We even have an Arabic (and Persian) instance of the rare mathematical-pedagogy genre of "lady's lectures" written by a prominent researcher. The eleventh-century polymath al-Bīrūnī composed his *Book on the Understanding of the Principles of the Art of the Stars*, an introductory work on astronomy and astrology, at the request of one Rayḥāna bint al-Ḥasan of Iran.

CHAPTER THREE

Inventing Mathematics

The Branches of Mathematics Take Shape

CLEMENCY MONTELLE AND KIM PLOFKER

Invention, or discovery, of new mathematical knowledge cannot be clearly separated from its professional practices, its technical applications, and its connections to other modes of thought. Mathematical research by advanced practitioners emerges organically from interest in particular types of problems and the sociocultural phenomena that provide incentive and impetus for understanding them. In turn, individual breakthroughs eventually reshape the standard mathematical background and training of practitioners. It is this perpetual "invention" of mathematics as a field of inquiry as well as a body of knowledge that this chapter explores in the context of the medieval world.

We begin with a brief overview of some shared features of academic mathematical traditions and taxonomic descriptions of their subject matter, as recorded around the middle to late first millennium by scholars writing chiefly in Latin and Greek, in Arabic and Persian, in Sanskrit, and in Chinese. We then survey a sample of the multitudinous significant developments that emerged in various areas of mathematical knowledge during the next several centuries. Although our modern subject classifications for these areas don't conform exactly to any of the traditions' own taxonomies, they reveal many connections between different mathematical cultures via shared or similar problems and techniques. Finally, we look at medieval discussions of foundational and methodological issues in mathematics, including the roles of demonstration, notation, and methods of discovery.

WHERE ARE MATHEMATICAL INVENTIONS FOUND?

A crucial aspect of all literate mathematics in this era, often insufficiently appreciated in the historical literature, was the prominent role played by oral communication. Not only instruction of pupils but the debates and discoveries of experts were carried out habitually through verbal discussion, and even written mathematics relied on words more than symbols. However, since written texts are the only part of medieval mathematical exposition that we can still directly access, they will be the focus of this chapter.

For readers who had progressed beyond the most elementary teaching, there was no clear dividing line in mathematical cultures between instructional textbooks and "research publications." Any work that contained material appropriate for the background and skills of a given student could be employed as a teaching text, while any treatise, canon, handbook, almanac, commentary, or correspondence that discussed a relevant subject could be an appropriate place to "publish" a new discovery. The following broad genres of scholarship were especially salient in the process of mathematical innovation.

Comprehensive Canonical Expositions

Among medieval adherents of Hellenistic learning, the ideal of a rationally organized exposition covering the entire discipline of mathematics and/or mathematical astronomy was most closely associated with the *Elements* of Euclid and the *Mathematical Syntaxis* or *Almagest* of Ptolemy. Arabic, Greek, and Latin recensions of these two texts (and the related "Middle Books") formed the core of these subjects in the Islamicate and Western worlds. In China, *The Ten Mathematical Classics* that students studied for government examinations supplied the basis of canonical knowledge. Indian mathematicians cited authoritative treatises or *siddhāntas* expounding many mathematical subjects within and outside astronomy, particularly the seminal works of Āryabhaṭa, Brahmagupta, Bhāskara I, and Bhāskara II.

Studies of Specialized Subjects

The genre boundaries between comprehensive canons and topic-specific "monographs" are likewise not always entirely clear. In this category might fall, for example, numerous Arabic works on arithmetic and algebra, such as al-Khayyāmī's *Treatise on Proofs of the Problems of Algebra* and al-Kāshī's *Keys of Calculation*. But entire books were also written on specific problems or techniques, like Nicholas of Cusa's works on circle squaring, Zhu Shijie's *Jade Mirror of the Four Elements* on equations with multiple unknowns, Nīlakaṇṭha of Kerala's *Computations Concerning Moon's Shadow*, or Abū Sahl al-Kūhī's analysis of the construction of the regular heptagon.

Commentaries

Probably the most important text genre in academic mathematical traditions, as in most other areas of medieval learning, was the scholarly commentary. Often distilled from lecture notes or other instructional materials into a systematic exegesis of a textual authority, a commentary could contain quotations from other works, sample problems illustrating a technique, or explanations of obscure passages. Outside the Euclidean sphere of influence, in which chains of mathematical statements were conventionally accompanied by formal demonstrations, commentaries were especially important for clarifying and justifying the rules in the base-text. A Chinese text comprising a sequence of briefly stated problems and solution methods, or an Indian one consisting of concise algorithms in Sanskrit verse, might be almost incomprehensible without an appropriate commentary. Many important commentaries in all traditions accumulated layers of subsequent "supercommentaries" expounding their original expositions.

Astronomical Tables and Manuals

Societal investment in astronomy, especially via government bureaus and royal patronage demanding revised astronomical models for improving calendars and timekeeping, provided mathematical practitioners with crucial support. It is not surprising that many interesting innovations in computational techniques, trigonometry, orbital modeling, and related areas appeared in astronomy works such as Islamicate *zījes*, Chinese calendar systems (*li*), and Sanskrit or Latin handbooks and tables.

EARLY MATHEMATICAL TAXONOMIES

Greek scholars in late antiquity famously partitioned mathematical knowledge into arithmetic, or the properties of numbers; geometry, or the properties of figures; harmonics, or arithmetic applied to music; and astronomy, or geometry applied to the heavenly spheres. Neoplatonist philosophers such as Iamblichus of Syria around 300 and Proclus at the Athenian Academy in the fifth century attributed this fourfold structure to the ancient Pythagoreans.

As Proclus explains in his commentary on Euclid:

> The Pythagoreans considered all mathematical science to be divided into four parts: one half they marked off as concerned with quantity, the other half with magnitude; and each of these they posited as twofold . . . Arithmetic, then, studies quantity as such, music the relations between quantities, geometry magnitude at rest, spherics [i.e., astronomy] magnitude inherently moving . . .
>
> But others, like [first-century BCE philosopher] Geminus, think that mathematics should be divided differently; they think of one part as

> concerned with intelligibles only and of another as working with perceptibles and in contact with them. By intelligibles, of course, they mean those objects that the soul arouses by herself and contemplates in separation from embodied forms. Of the mathematics that deals with intelligibles they posit arithmetic and geometry as the two primary and most authentic parts, while the mathematics that attends to sensibles contains six sciences: mechanics, astronomy, optics, geodesy [i.e., mensuration in general], canonics [applied harmonic theory], and calculation . . .
>
> Geometry in its turn is divided into plane geometry and stereometry . . . In the same way arithmetic is divided into the study of linear numbers, plane numbers, and solid numbers . . . Geodesy and calculation are analogous to these sciences, since they discourse not about intelligible but about sensible numbers and figures . . . Again optics and canonics are offshoots of geometry and arithmetic . . .
>
> Under mechanics also falls the science of equilibrium in general . . . and in general every art concerned with the moving of material things. There remains astronomy, which inquires into the cosmic motions, the sizes and shapes of the heavenly bodies, their illuminations and distances from the earth, and all such matters . . . The parts of astronomy are gnomonics, which occupies itself with marking the divisions of time by the placing of sun-dials; meteorology, which determines the different risings of the heavenly bodies and their distances from one another . . . and dioptrics, which fixes the position of the sun, moon, and stars by means of special instruments. (Morrow 1970: 29–35)

As noted in the Introduction, the basic fourfold division of mathematics into arithmetic, geometry, music, and astronomy survived as Boethius' well-known "quadrivium," which with the "trivium" of grammar, logic, and rhetoric constituted the seven "liberal arts" of medieval Latin learning. Under its Greek name *tetraktys*, the quadrivium likewise provided the standard framework for academic mathematics in Byzantium.

These classical categories, filtered through the early Islamic translations of Hellenistic sources, fundamentally influenced the conceptualization of mathematics in medieval Arabic scholarship and consequently in the Islamic world as a whole. The resulting Greco-Arabic corpus, despite endemic orthodox suspicions of its status as "pagan knowledge," became firmly entrenched in Islamic canons of learning. For example, the very influential tenth-century Arabic treatise *Enumeration of the Sciences* of the philosopher and music theorist al-Fārābī classified knowledge in general as comprising the linguistic sciences, logic, mathematics, physics, metaphysics, the civic sciences of ethics and political philosophy, law, and theology. Within that structure, mathematics was divided into the seven subjects arithmetic, geometry, optics, astronomy, music, statics,

FIGURE 3.1. Personifications of the seven liberal arts surrounding the throned figure of Philosophy along with Socrates and Plato, in a nineteenth-century reconstruction of a late-twelfth-century manuscript of excerpts from Scripture, philosophical and literary works, and other sources. This pictorial encyclopedia, entitled *Hortus deliciarum* (Garden of Delights), was compiled by the nun Herrad of Landsberg at Hohenburg Abbey in Alsace. Wikimedia Commons.

and technology, somewhat reminiscent of the eightfold classification of Proclus. Al-Fārābī's work was rendered into Latin in the twelfth century as *On the Sciences*, in versions by Gerard of Cremona and Dominicus Gundisalvi.

Another Arabic taxonomy of learning, the *Keys to the Sciences*, was composed in the late tenth century by the bureaucrat Muḥammad ibn Aḥmad al-Khwārizmī (not to be confused with the renowned ninth-century mathematician Muḥammad ibn Mūsa al-Khwārizmī). This compendium preserved the traditional Islamic distinction between the "Arabic sciences" (such as law, Islamic rationalist philosophy (*kalām*), and grammar) and the "foreign sciences" of "Greeks and other peoples." These latter subjects included *falsafa* (*philosophia*, Greek philosophy), logic, medicine, arithmetic, geometry, astronomy, music, mechanical devices, and alchemy, treated in accordance with Aristotelian criteria for distinguishing "theoretical" from "practical" knowledge. Of course, the importance of quantification was acknowledged within the Arabic sciences as well, particularly in al-Khwārizmī's chapter on *kitāb* "writing, scribal arts." As in the similar description of secretarial skillsets by Ibn Qutayba discussed in Chapter 2, al-Khwārizmī's bureaucratic mechanisms for collecting taxes, keeping accounts, and regulating official metrological and irrigation practices combined features from sources including the early Baghdad caliphal administration and their Sasanian predecessors. Numeracy played a role even in literary arts, such as prosody or analysis of poetic meters.

The organization of mathematical topics in medieval Chinese and Indian scholarship developed along different lines from their Greco-Roman and Greco-Islamic counterparts. The ancient Chinese classic *Nine Chapters* canonically structured mathematical knowledge by devoting each of its chapters to a different type of problem. These were categorized according to the methods by which they were to be solved, as follows:

1. Rectangular fields (plane geometry, determination of areas)
2. Grains (exchange rates computed using the Rule of Three)
3. Proportional distribution (division by proportional rates)
4. Small widths (dealing with fractions, extraction of square and cube roots, and determining the volumes of cubes and spheres)
5. Estimating workloads (problems of solid geometry, including determination of the volume of a square-based pyramid)
6. Equitable distribution (more complicated problems of proportional distribution)
7. Excess and deficiency (algebraic methods for solving excess/deficiency problems)
8. Rectangular arrays (algebraic methods for solving simultaneous linear equations)
9. Base and side (of right triangles) (problems devoted to properties of right triangles)

This classification scheme was retained as the basic framework of Chinese mathematics throughout the medieval period. Subsequent refinements and additions included Liu Hui's discussion of surveying geometry supplementing the ninth chapter on right triangles, and powerful new algebraic methods proposed by Song dynasty mathematicians.

The Sanskrit versified treatises and prose commentaries of medieval Indian *gaṇita* ("calculation, quantification, mathematics") invoked as their organizing principles various fundamental dichotomies that they perceived in their subject matter. For instance, the mathematician Bhāskara I in the early seventh century identified the following pairs of complementary concepts:

> This *gaṇita* is twofold . . . namely, increase and decrease . . . Increase is addition, decrease subtraction. All of *gaṇita* is comprehended by these two parts. [Some]one says: "Multiplication and [raising to] powers are parts of addition, and division and roots of powers [are parts] of subtraction . . ." Thus in the treatise (*śāstra*) and in the world, there is no method in *gaṇita* which is not either a form of increase or a form of decrease . . .
>
> Another says, "*gaṇita* is twofold, quantity and figure" . . . The *gaṇita*-kinds of proportions and indeterminate problems etc. are assigned to *gaṇita* of quantity; series and shadows etc. to *gaṇita* of figure. (Shukla 1976: 43–4)

Most subsequent works on *gaṇita* foregrounded the distinction between operations on known quantities (giving rise to the names *vyakta* "known," *pāṭī* "board," possibly a dust-board or other writing surface) and operations to solve for unknown ones (called *avyakta* "unknown," *bīja* "seed"). The former category embraced the fundamental "eight operations" (*parikarma*) of

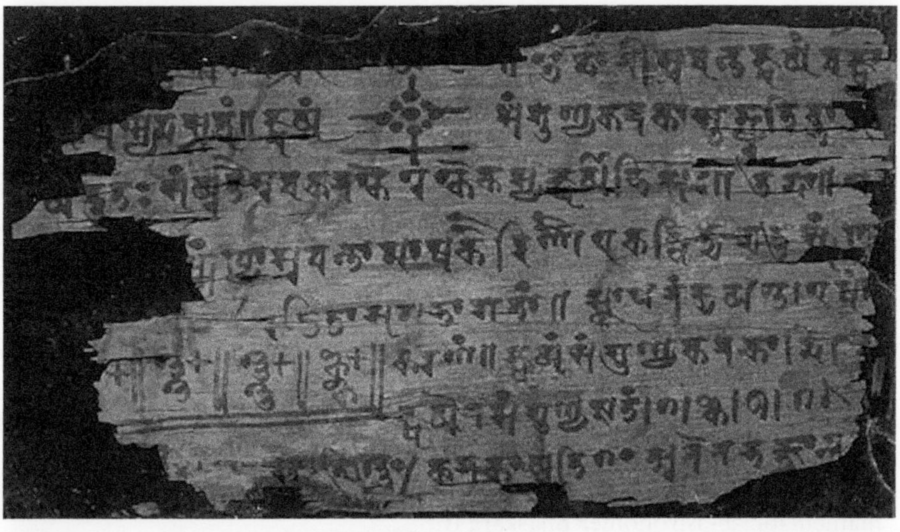

FIGURE 3.2. A page of the famous Bakhshālī Manuscript, written in Hybrid Sanskrit on birch bark sheets sometime in the first millennium, discovered in the late nineteenth century near Taxila (Pakistan). Mathematical expressions in boxed displays include vertically stacked fraction notation and negative signs in the form of small crosses (MS Sansk. D. 14, the Bodleian Library, University of Oxford, Oxford, UK). Photo courtesy of National Geographic / Wikimedia Commons.

arithmetic from addition to cube root, with their specialized forms for operating on fractional quantities; it also covered various "practices" (*vyavahāra*) or applications including mixture and proportion, series, geometry of plane figures, and computations involving solids and gnomon-shadow right triangles. The latter category included algebraic sign rules and techniques for solving determinate and indeterminate equations.

Some other topics, such as trigonometry and approximation techniques, were assigned by Indian scholars to the realm of astral sciences (Sanskrit *jyotiṣa*) rather than being treated in general mathematical texts. In fact, the genre of "general mathematical text" seems to have gradually supplanted an earlier classification of *gaṇita* as an auxiliary topic in *jyotiṣa*, in which astronomical treatises would devote a couple of chapters to non-astronomical computation including basic geometry formulas and calculating interest on loans.

ARITHMETIC OR COMPUTATION WITH KNOWN QUANTITIES

Elementary computational procedures (not to be confused with the number-theoretic concepts also frequently called "arithmetic") were the mainstay of the everyday mathematics discussed in Chapter 1, and also the gateway to "advanced" academic mathematics. Developments in practice and pedagogy substantially refined and expanded medieval arithmetic procedures including root extraction, computation with fractions, and solution formulas for problems in "mixture" or applied arithmetic.

The "Rule of Three"

Most prominent among these computational techniques was the renowned Rule of Three for problems in simple proportion. Despite its theoretical simplicity requiring only one multiplication and one division to find the desired result, the Rule of Three often lay at the heart of some very complex problem-solving in a wide range of contexts. It was employed in Chinese mathematical works as early as the pre-Qin or Qin text *Mathematics* written on bamboo slips no later than 212 BCE. By the mid-first millennium, it had been thoroughly analyzed in Indian mathematics, as attested by the earliest surviving Sanskrit arithmetic texts, where it already bears its standard name *trairāśika* "[rule] of three quantities." As was explained in the mid-twelfth century by the famous mathematician-astronomer Bhāskara II:

> Anything that is calculated in algebra or here [in arithmetic] by means of a multiplier and a divisor is understood by those of clear intelligence as just a Rule of Three Quantities. But for increasing the intelligence of dull-witted

> ones like us, it has been explained by the wise in many different and easy rules. (Plofker 2009: 186)
>
> Besides squaring, square-root, cubing, and cube-root, whatever is calculated is certainly a variation of the Rule of Three, nothing else. (Datta and Singh 1962: 210)

After being absorbed into Islamic mathematics along with other features of Indian arithmetic, the Rule of Three made its way to late medieval Europe, where it was subsequently nicknamed "the Golden Rule" or "the Merchant's key." The fifteenth-century German monk Friedrich Amman asserted in his *Practica des Algorismus Ratisbonensis* on commercial arithmetic:

> Therefore, first to be known is the rule of proportion, which among geometricians is called the Golden, among Italians the Rule of Three. It is the most useful of all rules included in arithmetic. The Rule of Three is also a proportional rule, of which always three numbers are given, but the fourth is unknown, and of which the proportion of the second to the first is known, but the proportion of the third to the fourth is unknown. (Berger 1990: 104)

Decimal Computation

We have briefly sketched in the Introduction the first decimal place-value numeral layouts using the ancient Chinese counting rods, and in Chapter 1 the spread and evolution of the Indian decimal-digit glyphs throughout medieval Islam and eventually Western Europe and Eastern Asia as well. This "Indian calculation," executed in ephemeral media such as the dust-board and slate as well as in ink on paper, gave rise to various computational formats and routines shared across different cultures.

One notable instance is the so-called "lattice multiplication" algorithm, first clearly attested in the thirteenth-century Arabic arithmetic work *Summary of Calculation Methods* of the Egyptian mathematician Ibn al-Banna' al-Marrākushī. He explains it thus:

> You construct a quadrilateral which you subdivide vertically and horizontally in as many strips as there are places in the numbers to be multiplied. Divide the squares [that are formed] by diagonals going from the lower left [corner] to the upper right [corner].
>
> Set out the multiplicand above the quadrilateral, making each of its places correspond to a column. Then set out the multiplier to the left or to the right of the quadrilateral [such that] it goes down it and also each place corresponds

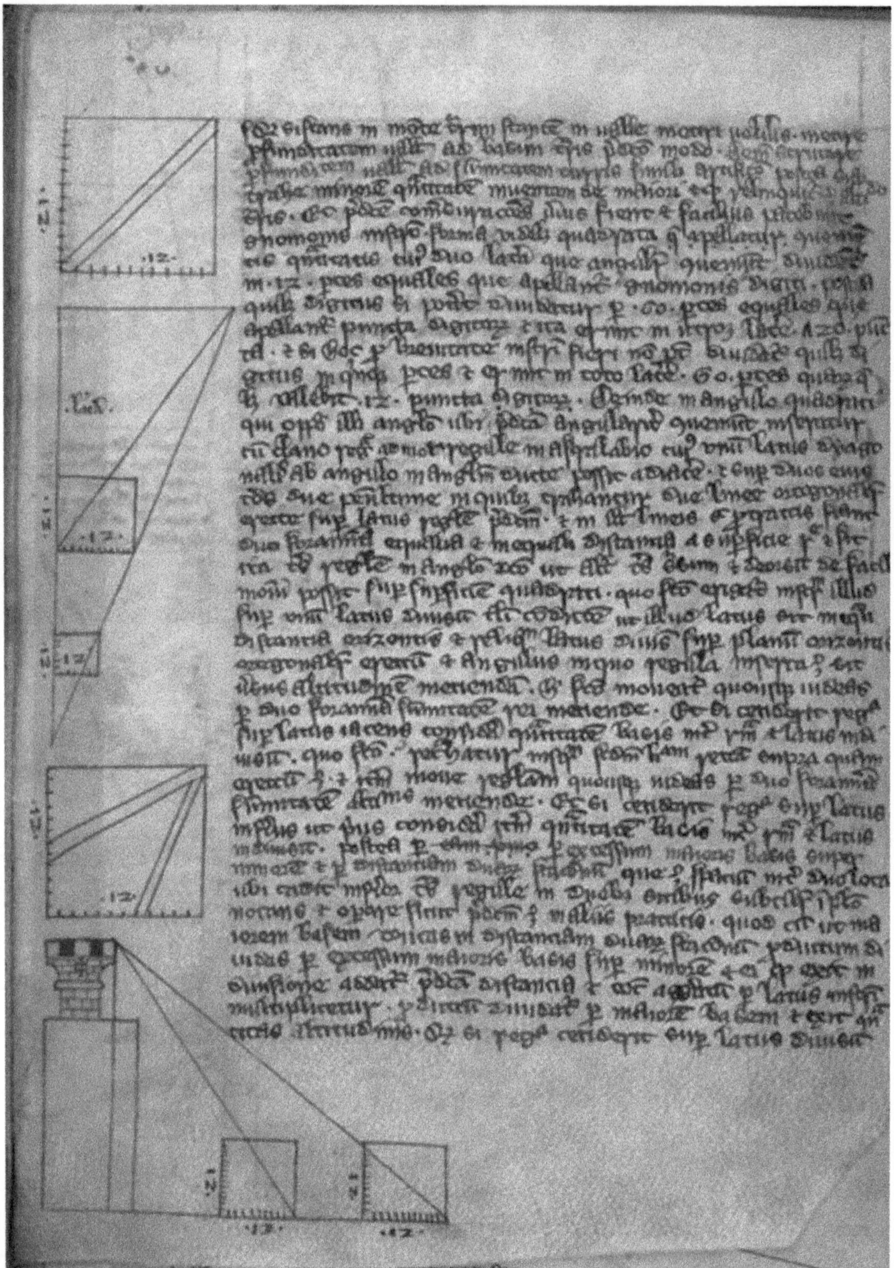

FIGURE 3.3. Diagrams on the geometric computation of altitudes and distances in a manuscript book containing a Latin treatise on mensuration and trigonometry, owned by an English university student around 1300. Courtesy of The Picture Art Collection / Alamy Stock Photo.

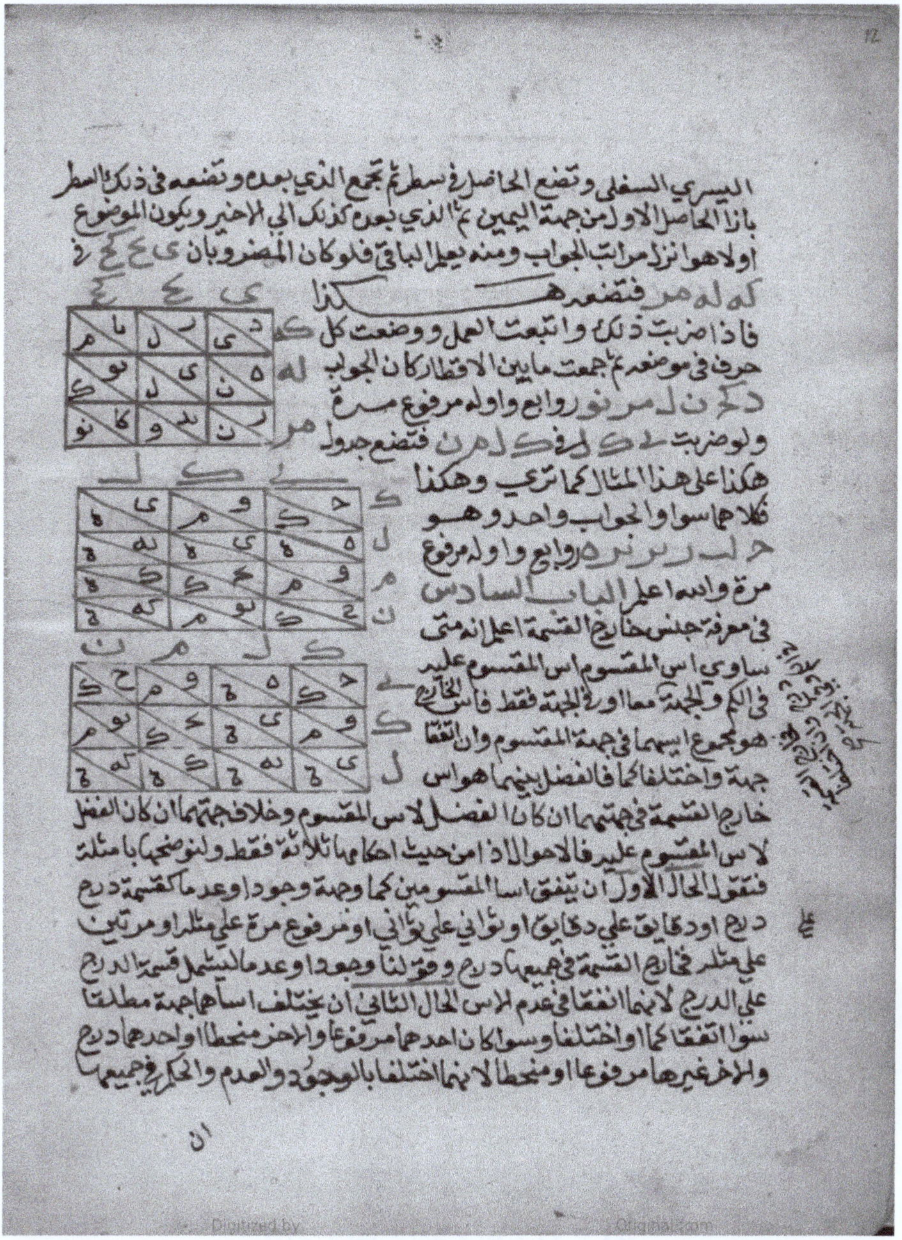

FIGURE 3.4. Lattice multiplication examples using Arabic alphabetic (*abjad*) numerals in a manuscript of an Arabic arithmetic treatise by the fifteenth-century astronomer Sibṭ al-Māridīnī. Courtesy of the University of Michigan and HathiTrust.

with a line. Then you multiply, one after the other, each of the places of the multiplicand of the square by all the places of the multiplier, and you put the [corresponding partial] product in each position in the square . . . (Chabert 1999: 21)

Called *shabaka* or *shubbāk*, "netting, window grille" in Arabic, lattice multiplication became popular in late medieval Europe chiefly under the synonymous Italian term *gelosia*, sometimes with artistically embellished grids. The technique was also referenced in India and in China at about the same time.

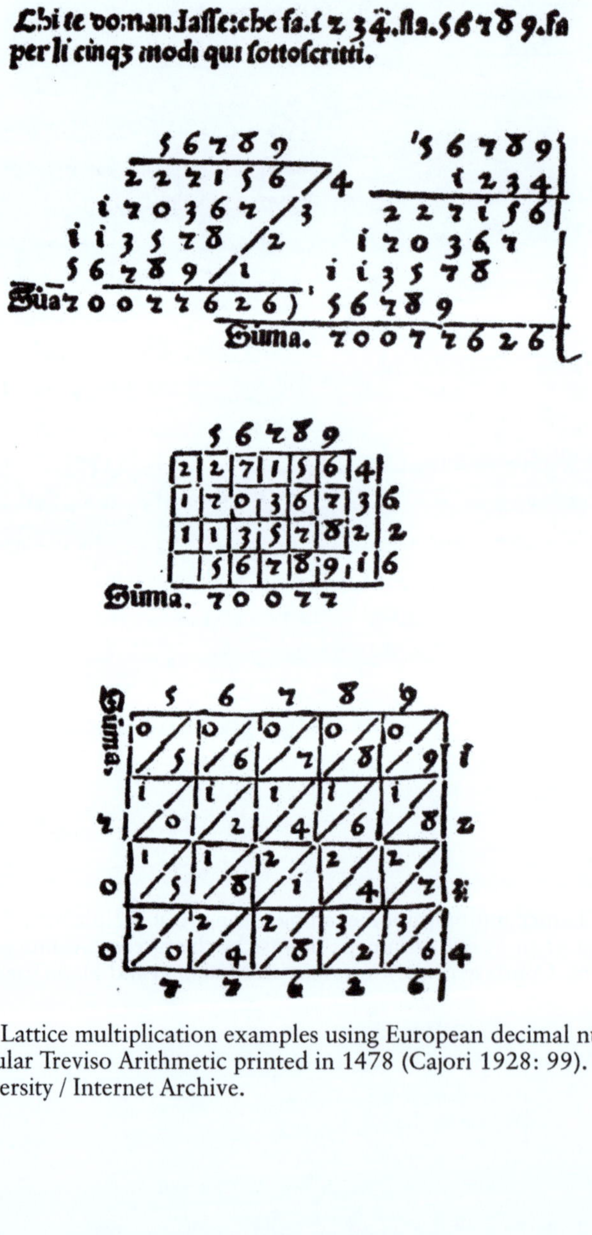

FIGURE 3.5. Lattice multiplication examples using European decimal numerals in the Italian-vernacular Treviso Arithmetic printed in 1478 (Cajori 1928: 99). Courtesy of Osmania University / Internet Archive.

ALGEBRA OR COMPUTATION WITH UNKNOWN QUANTITIES

This branch of mathematical knowledge includes various methods comprising what we now call algebra, linear algebra, indeterminate equations, etc. Elaborate classification schemes for equations and algorithms for solving them led to solutions for equations of higher order or in more than one variable, Diophantine problems, and systems of multiple equations involving subtle concepts of algebraic sign, manipulation of surd or irrational quantities, and so on. These ideas arrived in medieval European mathematics as a plethora of calculation rules for specific situations, from which Renaissance "cossists" and their successors gradually constructed a systematic symbolic approach to solving for unknown quantities. In so doing, European mathematicians were to some extent reinventing the systematized algebra of other cultures.

The Chinese "Elements"

As early as the third century in China when Liu Hui provided his commentary on the *Nine Chapters*, the use of algebraic matrix methods was already well established. The *fangcheng* method for solving systems of simultaneous linear equations in as many as five unknowns used an approach similar to Gaussian elimination, as in the following example:

> Now sell 2 cattle [and] 5 sheep, to buy 13 pigs. Surplus 1000 cash. Sell 3 cattle [and] 3 pigs to buy 9 sheep. [There is] exactly enough cash. Sell 6 sheep, [and] 8 pigs. Then buy 5 cattle. [There is] 600 coins deficit. Tell: what is the price of a cow, a sheep and a pig, respectively? (Shen 1999: 409)

The numbers of animals involved in each transaction are set out in a column, and entire columns are multiplied and subtracted in rectangular arrays to eliminate all but the desired unknown, whose value can then be determined.

By the time of the Southern Song around the turn of the second millennium, further developments in Chinese algebraic methods included the *tianyuan* ("heavenly element") method for handling root extractions and solutions of polynomial equations of any order, as well as the "four elements" method used to solve equations with as many as four unknowns. Both of these, like the earlier matrix-type techniques with rectangular arrays, used specific locations on the counting board designating various powers and combinations of unknowns to set up equations that could then be solved by carefully worked out algorithmic procedures reducing them to simpler equations.

These standardized layouts included early forms of the binomial-coefficient grids now commonly called the "arithmetical triangle" or "Pascal's triangle,"

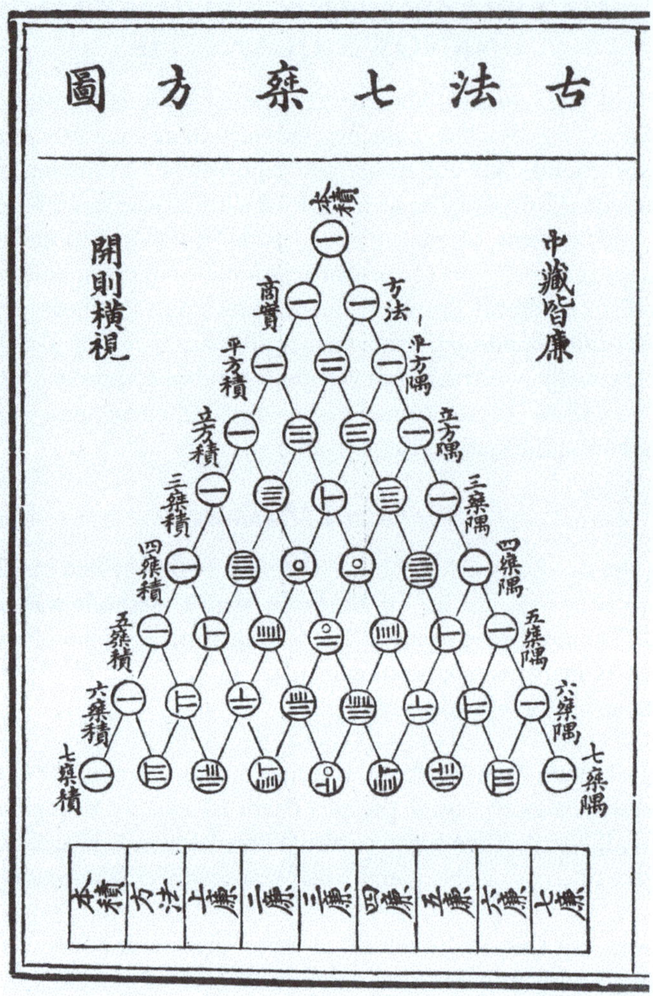

FIGURE 3.6. The binomial-coefficient "arithmetical triangle" constructed by Yang Hui in his thirteenth-century commentary on the *Nine Chapters*, as preserved in the *Yongle Encyclopedia*. In his book, Yang Hui noted that his method for finding square and cube roots using "Yang Hui's Triangle" was in fact derived from the work of Jia Xian, a Northern Song dynasty mathematician (Needham 1959: 135). Courtesy of the Needham Research Institute, Cambridge, UK.

used to facilitate root extractions and solutions of polynomial equations. A version presented by Zhu Shijie in his 1303 *Jade Mirror of the Four Elements* includes coefficients up to the eighth power. A sixth-power version had been described earlier in the 1261 *Detailed Explanations of the Nine Chapters* of Yang Hui, and attributed by him to the mid-eleventh-century author Jia Xian.

The Indian "So Much"

In India, too, the solution of problems involving one or more unknown quantities (generally denoted by Sanskrit *yāvattāvat* "as much as so much," abbreviated *yā*) had attained by the mid-first millennium a high degree of structure and sophistication. Bhāskara II in the twelfth century asserted the systematic power of *bīja-gaṇita* or algebra rules that could replace mathematical intuition for the "slow-witted":

> Algebra, [which is] equivalent to the rules of arithmetic, appears obscure, but it is not obscure to the intelligent; and it is [done] not in six ways, but in many . . .
>
> Previously mentioned [in the *Līlāvatī*] was the manifest [arithmetic] whose source is the unmanifest [algebra]. Since generally questions cannot be very well understood by the slow-witted without the application of the unmanifest, therefore I tell also the operation of algebra. (Plofker 2009: 192)

(The "six ways" refers to the subset of the eight fundamental arithmetic operations, omitting cubing and cube-root, that were also employed in algebra.)

Bhāskara in his algebra text *Bījagaṇita* comprehensively discussed the solution of linear and quadratic equations or *samī-karaṇa,* literally "making equal." In accordance with Indian algebra notation conventions, the "equals" were written out as vertically stacked paired expressions combining numerical coefficients with verbal abbreviations for unknowns. Using types of completing-the-square techniques for solving quadratic problems, Bhāskara extended them to certain cubic and quartic equations as well. Moreover, he explicitly articulated some constraints on quadratic solutions, such as the condition for obtaining two positive roots.

Indian algebra also addressed indeterminate equations, particularly the first-degree type now known as linear congruences. Indian techniques for such problems are first attested in an astronomy treatise by Āryabhaṭa around 500; they seem to have originated in early Common Era mathematical-astronomy models that required synchronizing integer numbers of revolutions completed by celestial bodies in immense integer numbers of years. The models' parameters were deduced by solving linear indeterminate equations that went by the name of *kuṭṭaka* or "pulverizer/grinder," as their solution procedure (equivalent to the Euclidean algorithm) involved "grinding down" a given indeterminate equation to one with smaller coefficients.

Second-degree indeterminate equations expressing relations between given numerical constants and two squared unknowns also captured the imagination of many Indian algebraists, for historical reasons less easily ascertained. The work of Brahmagupta in the seventh century is the earliest extant Sanskrit source of these "square-nature" (*varga-prakṛti*) problems, similar in scope to the so-called "Pell's equation" investigated by early modern European mathematicians. Brahmagupta's approach generated solutions to a given square-nature problem by composing solutions found for simpler forms of it. Within the next few centuries, a brilliant alternative technique had been proposed, called "cyclic" (*cakravāla*), which generated the complete integer solution.

The Arabic "Thing"

The profound impact of decimal place-value "Indian calculation" on arithmetic methods in the Islamic world did not lead to a similar adoption of Indian algebra. Traditions in Islamic algebra seem to have taken their inspiration primarily from "subscientific" practical algorithms with an admixture of Euclidean and Diophantine abstraction. A seminal Arabic work on the science of solving equations with unknown quantities was the previously mentioned *Algebra* or *Book on Calculation by Restoration and Confrontation* composed in the first half of the ninth century by the Persian mathematician Muḥammad ibn Mūsa al-Khwārizmī. This work, apparently drawing on a body of algebra knowledge already quite mathematically mature, primarily addressed problems arising from inheritance distribution, commercial transactions, and land mensuration.

One of the features of al-Khwārizmī's text that became standard in later algebra works was its classification of particular types of equations, to which all problems involving unknowns of the first and second order could be reduced. It distinguished three types of elements: constant numbers (often designated by the currency unit *dirhams*), the "thing" (i.e., the unknown, Arabic *shay'*) or "root," and the "square" (i.e., the square of the unknown, Arabic *māl*). From these three elements were derived six basic types of equations to which all such problems could be reduced, expressed with the following rhetorical labels rather than with the symbolic notation of their accompanying modern equivalents:

Al-Khwārizmī's expression	*Modern equivalent*
squares equal roots	$ax^2 = bx$
squares equal numbers	$ax^2 = c$
roots equal numbers	$bx = c$
squares and roots equal numbers	$ax^2 + bx = c$
squares and numbers equal roots	$ax^2 + c = bx$
squares equal roots and numbers	$ax^2 = bx + c$

The approach to solving the example problems for each of these cases is twofold, with a numerical solution preceding a geometric derivation whose details sometimes require at least five times as much exposition as the preceding algebra procedure. A generation later, mathematicians such as Thābit ibn Qurra and Abū Kāmil justified the solutions to these and more complicated problems (including some involving irrational square roots) in more detail with recourse to geometric propositions from Euclid's *Elements*. Abū Kāmil's synthesis also offered proofs of some geometric propositions by means of algebra. Other authors explained alternative problem-solving techniques such as single or double false position along with algebra procedures.

Qusṭā ibn Luqā's Arabic translation of Diophantus a few decades after al-Khwārizmī's *Algebra* was the earliest extant *al-jabr* work to incorporate the third power of the unknown (named "cube") along with higher powers denoted by compound terms such as "square of the square," "square of the cube," etc. In the work of al-Karajī in the eleventh century, reciprocal powers (what we would call negative powers) of unknowns were recognized, along with determinations of binomial coefficients employing a form of the well-known arithmetical triangle, which was also used in the twelfth-century algebra of al-Samaw'al. The twelfth century likewise brought forth (especially owing to the contributions of

FIGURE 3.7. The binomial-coefficient "arithmetical triangle" using Arabic forms of decimal place-value numerals, in a manuscript of the twelfth-century Arabic treatise *Al-bāhir fī al-jabr* (The Splendid Book on Algebra) by the mathematician al-Samaw'al (MS Suleymaniye Aya Sofya 2718, fol. 51v). Courtesy of Süleymaniye Manuscript Library, Istanbul, Turkey.

'Umar al-Khayyāmī or Omar Khayyam) the classification of equation types extended to cubic equations, along with the use of intersecting conic sections to solve irreducible cubics. Although mathematicians in the Islamic East such as the Iranians al-Karajī and al-Khayyāmī held to the purely rhetorical style of algebra exposition, some of their contemporaries in North Africa developed a symbolic notation with superscript and subscript abbreviations.

For the most part, it was the more elementary material of Islamic algebra that was transmitted to Western Europe through efforts like the Toledo translation project mentioned in Chapter 2. In particular, Gerard of Cremona's Latin translation of al-Khwārizmī's work on what Europeans called "Algebra and Almuchabala" laid the foundations of algebra instruction in Europe. A similarly titled section in Leonardo of Pisa's 1202 *Book of Calculating* testifies to the continuing interest in this subject, which became the "cossick art" (from the Italian rendering *cossa* or *cosa* of Arabic *shay'* "thing") of Renaissance algebraists.

Operations with Zero and Negative Quantities

Arithmetic algorithms for operating with known quantities could be readily extended beyond the natural numbers, leading to questions such as whether a numerical quantity could be literally nothing, or even less than nothing. These questions were answered differently in different traditions. For example, well before the start of the medieval period, Chinese counting-rod arithmetic had evolved representations for both positive and negative quantities: sometimes using black-colored rods for the former and red for the latter, and sometimes placing a slanted rod across the number to be considered negative (a notation easier to represent in monochrome printed diagrams of counting-rod layouts). Sign rules were also enunciated to handle the interactions of positive and negative, as well as stating what to do when adding or subtracting a positive or negative number from zero, although a zero digit in a place-value number was denoted not by a specific symbol but by an empty place.

In contrast, medieval Indian mathematics recognized zero as a number, albeit one requiring some special arithmetic rules, and employed it as an operand in expositions such as the following from Bhāskara's twelfth-century arithmetic text *Līlāvatī*:

> In addition, zero [produces a result] equal to the added [quantity], in squaring and so on [it produces] zero. A quantity divided by zero is "zero-divided" (*khahāra*); [a quantity] multiplied by zero is zero, and when there are remaining operations, it is considered as "zero-multiplied" (*khaguṇa*). When [a number] is multiplied by zero, if it is also divided by zero, then the quantity is to be understood as unchanged (*avikṛta*), and similarly when it is decreased and increased by zero. (Plofker 2009: 185)

This approach to division by zero, which to modern readers seems naïve and unclear, may have been a deliberately simplified version of the concept for beginning arithmetic students. Compare, for example, the more nuanced description by Gaṇeśa in the sixteenth century in his commentary on the *Līlāvatī*:

> A quantity which has attained zero status when multiplied by zero, will not get its original status back if divided again by zero. It is just like a yogi, who by his spiritual practice realises the Brahman-nature, which is without a second, untainted, and of the nature of consciousness and bliss, [and] will not return to his earlier status of bondage. (Ramakalyani 2016: 53)

Negative quantities were another standard concept in Indian mathematics, with formalized sign rules for their operations. They were frequently denoted by a special symbol, such as a dot above them or a small cross following them. This aspect of "Indian calculation" was less readily accepted by mathematicians in the Islamic world. Hints of the confusion in perceptions of negative numbers within European mathematics can be seen, for example, in some scholia in early ninth-century Latin mathematical manuscripts grappling with their rules:

> True with true makes true. Less with true makes true. True with less makes less. Less with less makes less. True signifies being; less means nothing. Put any sum of numbers you wish called true, that is being, and put another sum you wish called less by the adverb that I said signified nothing, and combine the two sums. That which was larger conquers the less and consumes it according to the quantity of its size. (Katz 2016: 11–12)

Leonardo of Pisa's abovementioned *Book of Calculating* includes some problems using what are effectively negative quantities of money treated as debts, but without a generalized arithmetic of positives and negatives. That capacity emerged in European mathematics with the post-medieval elaboration of algebra.

NUMBER THEORY, SERIES, COMBINATORICS

In addition to the operations of arithmetic and algebra, medieval mathematicians studied several other aspects of number. For instance, concepts from classical number theory like integer primality and amicability were explored among the Islamic and European followers of the Euclidean tradition. The number of possible ways to select choices from a set of options was explored in contexts from identifying metrical patterns in Sanskrit verse, to enumerating all the possible Arabic words formed from a given number of root consonants, to counting planetary conjunctions.

Arranging and sorting elements was a popular theme in early Chinese divination traditions, including the interpretation of glyphs composed from arrangements of lines. These were the mystical sixty-four "hexagrams" consisting of patterns of six line segments, each of which might be broken (representing the cosmic principle *yin*) or unbroken (representing *yang*), used for divining as far back as the *Book of Changes* in the first millennium BCE. Chinese mathematicians considered not only the possible variations of the figures, but also logical ways to order them. Later algorithms for predicting the course of illness could include arithmetic manipulations based on numerical inputs such as the patient's age and the month and day numbers of the onset of illness.

Games involving cards, dominoes, dice, and other types of playing pieces also provoked interest in combinatorial variation, illustrated by the eleventh-century polymath Shen Gua's analysis of the configurations in the game of Go. The ancient cosmic significance of magic squares, and techniques for their construction, offered another pathway into studying the arrangement of integers. A magic square of order three is attested as early as the Han dynasty and ascribed to even earlier ancient representations of the "Luo River Diagram" (or "Nine Halls Diagram"), said to have been inscribed on the shell of a magic turtle that emerged from the Luo River. Magic squares in general were considered to be endowed with mystical, magical properties: architects used them in the design of tombs, temples, and even for city planning. The magic square also has affiliations with the numerology of the *Book of Changes* and the geomancy of *feng shui*.

By the time the mathematician Yang Hui introduced squares of order greater than three in his late thirteenth-century *Continuation of Ancient Mathematical Methods for Elucidating the Strange* [*Properties of Numbers*], their divinatory function seems to have given way to purely arithmetical investigation. Yang Hui considers in all a total of twenty magic squares and circles, one of which of order nine is made up of nine magic squares of the third order; another is a "semimagic" square of order ten, i.e., one whose diagonals do not have the same sum as the rows and columns.

Techniques of combinatorial analysis in India likewise went back to antiquity, first attested in texts on Sanskrit prosody (*chandas*) explaining how to determine the possible arrangements of light and heavy syllables in a verse meter of a given length. In medieval *gaṇita* texts the enumeration and ordering of combinations and permutations was applied to problems such as selecting musical notes for a melody, flowers for a garland, or tastes or fragrances for a medicine or perfume. Standardized diagrams were developed to assist in these computations, including

FIGURE 3.8. An Indian magic square of order four with sum 34, engraved in the early second millennium on the Parshvanatha temple in Khajuraho, Madhya Pradesh. Photo courtesy of Kim Plofker.

a version of the arithmetical triangle named after the sacred cosmic mountain Meru. The most comprehensive treatment of combinatorial problems, which had acquired the Sanskrit name "net of digits" (*aṅkapāśa*), was that of Nārāyaṇa Paṇḍita in the mid-fourteenth century in a work that also collected and analyzed numerous techniques for magic square construction.

Abū Rayhān al-Bīrūnī described some features of Sanskrit prosodic combinatorics in his Arabic survey of Indian culture and science. Elementary enumerations of combinations in Arabic linguistics, music theory, medicine, and other contexts would have been already familiar to many of his readers. So might some of the mathematical aspects of magic squares, whose auspicious properties made them popular as amulets and talismans.

Around the thirteenth century, some mathematicians in the Islamic Maghreb explored algorithms to express the number of combinations and arrangements of a subset of a given number of letters, along with formal proofs or rationales for these algorithms. Among them, Ibn Mun'im created an explicit formula for expressing permutations of letters without repetitions—which he depicted with the help of what appears to be the first purely combinatorial construction of an arithmetical triangle—while Ibn al-Bannā first enunciated the standard formula for generating binomial coefficients. The latter also applied some of his combinatorial reasoning to other mathematical topics including sums of arithmetic progressions.

Hellenistic interest in number-theoretic properties of integers was carried forward by Islamicate mathematicians such as Thābit Ibn Qurra, author of a treatise on amicable numbers, and ibn Mun'im who wrote extensively on figurate numbers. In European settings, of course, the fundamentals of this subject—even and odd, primality, perfect numbers and related concepts, polygonal and other figurate numbers, ratios and means—constituted the "arithmetic" part of the standard quadrivium, as well as most of the "music" part dealing with harmonics. The "Pythagorean" mysticism of sacred numerology among Christians and Jews added to the perceived prestige of such number properties, extolled in works like the play *Wisdom* of the tenth-century canoness Hrotswitha of Gandersheim.

The thirteenth-century Catalan astrologer and philosopher Ramon Llull sought to automate the power of numerological insights (and the pursuit of logical discourse in general) with a simple combinatorial algorithmic device or "art" applied to sets of abstract principles such as goodness, greatness, eternity, and so forth. Number and numerology were likewise blended in the magic-square diagrams that Europeans apparently first learned of through the Jewish translator and polymath Abraham ibn Ezra. The Byzantine scholar Manuel Moschopoulos in the fourteenth century wrote a treatise on magic squares, also probably inspired by Islamic sources.

GEOMETRY OR PROPERTIES OF FIGURES

Ancient mathematics provided the foundations for innumerable medieval explorations and demonstrations of the nature of geometric shapes. These shapes might model real-world phenomena, like cyclic quadrilaterals inscribed in circles of planetary orbits, or polygons representing quantities such as distance or time in the late medieval scholastic discourse of "latitude of forms." Or they might be highly abstract refinements of earlier results, such as constructions of the regular heptagon and similar findings in the Islamicate Euclidean tradition. Geometric arguments were also increasingly used for verification of results in other areas of mathematics, especially algebra.

Chinese geometers writing detailed supercommentaries upon Liu Hui's presentation of the *Nine Chapters* expanded many of the results and arguments he recorded there. For example, Liu Hui's analysis of circle-area formulas by means of polygons with increasing numbers of sides presumably inspired Zu Chongzhi's computation of the notably accurate circumference-diameter ratio value of 355/113 near the end of the fifth century. *The Ten Mathematical Classics* compiled in the seventh century likewise extended Liu Hui's stereometric dissection techniques for computing polyhedral solid volumes to rationales for volumes of non-polyhedral shapes such as the cone, cylinder, conical frustrum, and sphere. The justification given by Zu Xuan, the son of Zu Chongzhi, for the volume of the so-called "double vault" solid (equivalent to Archimedes' bicylinder) is perhaps the most sophisticated of these, invoking what might be considered a three-dimensional form of Cavalieri's principle.

The ancient Chinese "out-in" principle, in which equality of non-congruent areas is demonstrated by subdividing figures so that identical parts lie "outside" and "inside" the boundary, remained the main tool for problems involving right-triangle geometry. Related dissections were used to determine the dimensions of circles or squares inscribed in right triangles. However, the application of geometric dissection most popularly associated with Chinese sources—namely, the type of seven-piece puzzles widely known as "tangrams"—appears to be an early seventeenth-century innovation which remained a recreational puzzle rather than an object of theoretical study.

Collections of standard formulas for computing the dimensions of various figures formed the core of geometry in Indian *gaṇita* texts. As mentioned in Chapter 2, some of these rules were explicitly connected to "real-world" problems involving heaps of grain or stacks of bricks, while others dealt with more abstract figurations. But as in Chinese mathematics, deeper explorations of geometric reasoning emerged in many of the commentarial demonstrations of the formulas, and in geometric interpretations of other types of mathematical problems. Instead of a structured sequence of axioms, propositions, and constructions, this reasoning relied on the reader's ability to grasp less formal rhetorical arguments (such as Bhāskara II's common-sense explanation of the impossibility of a rectilinear figure having the sum of all but one of its sides equal to or less than the remaining side).

Highlights of this tradition include the following: results for the area, diagonals, circumradius, and construction with rational side lengths in the case of a quadrilateral inscribed in a circle ("cyclic quadrilateral"), especially as discussed in the works of Brahmagupta and Nārāyaṇa Paṇḍita; values of the circumference-diameter ratio including the ancient $\sqrt{10}$ attested in Jain works and Āryabhaṭa's value of 62832/20000; and geometric rationales for formulas for, e.g., the area of a circle and the volume of a sphere. Verbal descriptions of manipulating squares and rectangles were sometimes employed to verify algebra results, including indeterminate equations. Geometric concepts might also serve in the demonstration of statements about series, like Nīlakaṇṭha's rationale for the sum formula of the first n triangular numbers as an arrangement of rectangular prisms representing the successive terms.

West of India, the long shadow of Euclid and his successors greatly influenced perceptions of the nature and aims of geometry. Translations and commentaries on not only Euclid's own works but those of Archimedes, Apollonius, Diophantus, Ptolemy, Pappus, and other Hellenistic scholars laid the groundwork for geometric research in the Islamicate world. The foundations of Euclidean geometry including the parallel postulate were investigated by, among others, al-Khayyāmī and al-Ṭūsī. Geometric proofs in algebra, constructions of regular polygons and conic sections, inscribed and circumscribed figures, locus problems, centers of gravity, proportions, and many more topics gave rise to new approaches and discoveries featured in treatises and commentaries by authors too numerous to list.

Optics, too, was extended from its Euclidean roots to a systematic study of the geometry of light and vision, whose main contributors included the ninth-century polymath al-Kindī and especially Ibn al-Haytham. The design of architectural form and ornament, particularly the elaborate vaults called *muqarnas* and complex patterns of polygonal tiles, provided further scope for geometers' virtuosity. (For examples of these, see Figure 6.4.)

Applications of geometry in the Latin world received little academic attention until the late medieval "practical geometry" compilations mentioned in the Introduction. Geometric theory likewise persisted in the quadrivium of the first millennium at what has been described as a "sub-Euclidean" level, presenting definitions and elementary propositions without demonstrations (along with some excerpts from *agrimensores* literature). As discussed in previous chapters, the early second millennium loosed a flood of Latin recensions of Greek and Arabic geometry texts. By the early fourteenth century, innovative studies of Euclid appeared from, e.g., the French Jewish philosopher Levi ben Gerson and the scholastic and theologian Barlaam of Seminara in Constantinople. The ideas of High Scholastic writers like Bradwardine and Oresme on the application of simple geometry to the quantification of motion and change have already been touched upon in the Introduction.

TRIGONOMETRY OF TRIANGLES AND CIRCLES

Sharing a common ancestor in Hellenistic Greek trigonometry of chords, various methods for quantitatively relating triangle sides to angles and/or circular arcs developed in various strands of medieval mathematics together with geometric study of circles and triangles. (The succession of cross-linguistic loanwords that produced the modern name "sine" is described in the Introduction.) Their connection to Hellenistic spherical astronomy (explored further in Chapters 4 and 5) is reflected in their 360-degree circles, in their accompanying celestial coordinate systems, and in their focus on the mathematization of space.

Some of the earliest known Indian mathematical astronomical works involving trigonometry likewise show traces of the subject's Hellenistic context: e.g., features of horoscopic astrology, including the twelve zodiacal signs with their Greek names, the seven-day week, and sexagesimally notated trigonometric values. Like its Hellenistic forebear, Indian plane trigonometry dealt with lengths of right triangle sides scaled to a standard non-unity value of the hypotenuse, rather than with dimensionless ratios between their lengths.

By the time the oldest extant forms of these Sanskrit works were compiled in the mid-first century, Indian mathematicians had already modified the original trigonometric chord or "bowstring" (Sanskrit *jyā* or *jīvā*) of a given arc or "bow" to the "half-bowstring" or half-chord of double the arc. This new half-chord quantity, which eventually superseded the original full chord as the referent of the terms meaning "bowstring," was presumably preferred on account of the increased convenience it afforded for table lookup. (That is, an angle in a right triangle inscribed in a semicircle must be doubled in order to find the corresponding chord, whereas the same angle in a right triangle with hypotenuse equal to the circle radius corresponds directly to the tabulated opposite side or half-chord.) The most commonly employed Indian value for the trigonometric radius was 3438 (or about $360 \times 60/2\pi$) of the units equated to the arclength of an arcminute, a convention resembling modern radian measure.

Besides the half-bowstring or sine and its variant form of complement-sine or *koṭi-jyā*, the other standard tabulated trigonometric quantity was the so-called "arrow" (*śara*) or versed sine (*utkrama-jyā*), equal to the difference between the radius and the complement-sine. Relations involving plane triangles inside the sphere, or sometimes approximating a small triangle on its surface, were ramified into innumerable formulas for determining astronomical quantities, often employing the similarity of a trigonometric *jyā* and *koṭi-jyā* to a (vertical) gnomon and its (horizontal) shadow, respectively. More general trigonometric results were also derived, such as the formula for the sine of the sum of two arcs recorded in the twelfth-century "Sine-generation" (*Jyotpatti*) attributed to Bhāskara. The culmination of Indian trigonometry is often taken to be the detailed demonstrative study of the subject preserved in the works of the school of Mādhava in Kerala from the fourteenth through sixteenth centuries, which included the power series expansions discussed in a later section of this chapter.

Islamic mathematical sciences absorbed both Hellenistic chords (particularly in spherical trigonometry) and Indian sines, which seem to have been known to Arabic authors from at least the late eighth century via a translated Sanskrit astronomical work called *Sindhind* in Arabic. Along with the Indian sine, Islamic mathematicians standardized other functions in their trigonometric tables, which were typically compiled for every degree of arc and scaled to a radius of 60. By the tenth century, these included tables of the versine (*sahm* "arrow" in Arabic), tangent and cotangent ("reversed shadow" and "direct shadow" respectively), and secant and cosecant (the corresponding "hypotenuse" of each shadow). "Shadow square" scales were engraved on the backs of astrolabes to facilitate the use of the "direct shadow" and "reversed shadow" relations in measuring the height of objects. Many auxiliary functions expressing astronomical quantities were also tabulated in *zījes*, including declination and rising-time; they were applied in various astronomical and geographical problems such as the crucial determination of the *qibla*.

Spherical triangle relations such as the law of sines were articulated over the next few centuries in works by mathematicians like al-Bīrūnī, Abū al-Wafā', and al-Ṭūsī, while computational refinements allowed function values to be tabulated at intervals as fine as one arcminute. The sophistication of this trigonometric knowledge fostered the production of detailed mathematical projections from the sphere to the plane to construct various kinds of astrolabes, analemmata, and sundials. Beginning in the late fourteenth century, some of these developments in Islamic trigonometry were in turn assimilated into its Indian counterpart via Sanskrit versions of Persian treatises, including some works on the astrolabe and some *zījes* or astronomical tables.

The Islamic *zīj*, astrolabe, and related astronomical apparatus also helped to introduce trigonometry to medieval Europe. Byzantine learning preserved Ptolemy's mathematical astronomy in Greek, and such thirteenth- and fourteenth-century scholars as Theodore Metochites and Nicephorus Gregoras undertook some renewed theoretical study of it. However, the main influence on late medieval Greek astronomy (and astrology) was the translation of various Persian *zījes*. The Greek *zīj* translation made around 1300 by Gregory Chioniades after long study in Tabriz with the astronomer Shams al-Bukhārī called the sine *trachelaia* (possibly a painfully literal translation of Arabic/ Persian *jayb* as "neck-opening"?) and the versine *sagita*, presumably from the Latin for "arrow." Chioniades confined his investigation of these quantities to explaining the use of their tables in calendric and planetary computations. (His familiarity with this material was apparently considered sufficiently suspect that around 1305 he had to repudiate publicly any belief in astrological divination, considered by the Patriarchate to be a heterodox superstition.)

The early twelfth-century translation of the ninth-century *zīj* of al-Khwārizmī by Robert of Chester, who also translated the same author's more famous work on algebra, contains the first known rendering of the Arabic term *jayb* into Latin *sinus*. Other late medieval Latin authors described some trigonometric concepts, including Leonardo of Pisa in his 1220 *Techniques of Geometry* which includes a chord table with stepsize equivalent to 30 degrees and chords measured in rods, feet, inches, and points. About a century later, the Latin translation of an astronomy treatise contained in Levi ben Gerson's massive Hebrew philosophical work *Wars of the Lord* included a geometric exposition of basic sine-and-versine trigonometry entitled *On Sines, Chords and Arcs*. His contemporary Richard of Wallingford's *Four Books on Sine Demonstrations* likewise systematically explained and proved various propositions about the geometric relations of the *sinus rectus* or sine, the *sagitta* or *sinus versus*/versine, and the *sinus duplatus* or *corda duplata*, the "double sine" or chord. Trigonometric features percolated into vernacular literature as well, e.g., in the description of the *umbra versa* and *umbra recta* shadow scales in Geoffrey Chaucer's late fourteenth-century English treatise on the astrolabe.

The revival of interest in the theory of Ptolemaic spherical astronomy was further stimulated by the new Latin translation and commentary on the *Almagest* produced by George of Trebizond for Pope Nicholas V in 1451. The comprehensive reworking of Latin trigonometry by Regiomontanus and his successors discussed in Volume 3, including the invention of the name *trigonometria*, was ultimately inspired by this astronomical renaissance. As Regiomontanus remarked in his 1464 *De triangulis omnimodis*, "The science of the stars will not be sufficiently understood by anyone neglecting our triangles."

In the meantime, Indian trigonometry based on the Hellenistic 360-degree circle had entered the ambit of Chinese calendrics via translated Sanskrit works as early as the eighth century, but its influence on Chinese mathematics and mathematical astronomy was minimal. (For one thing, the circle in Chinese astronomy was divided into 365 1/4 parts, not readily compatible with standard trigonometric degrees.) Indian sine values were incorporated into some Tang-era astronomical formulas for quantities such as lunar latitude and solar longitude, and in the thirteenth and fourteenth centuries Islamic *zījes* with their trigonometric tables were used by Muslim astronomers at the Islamic Astronomical Bureau at the Yuan and Ming courts. It was the exposition of early modern Western trigonometry by Jesuit scholars a few centuries later that launched more general Chinese investigations of the relations of right triangle sides and angles.

APPROXIMATION, INTERPOLATION, ITERATION

Approximate numerical values were frequently employed for constants that could not be expressed as integers or simple fractions. However, it is seldom that a clear path of transmission can be traced between appearances of the same value in different mathematical traditions. Of course, the ancient estimate of 3 for the circumference-diameter ratio now called π was used all over the medieval world. The more accurate π values 22/7, $\sqrt{10}$, and various fractions equal to 3.1416 are all mentioned in Chinese, Indian, and Islamic texts; al-Khwārizmī may have found the latter two in Indian sources. Traditional approximations of "surds" or square roots of non-square numbers included 7/5 for $\sqrt{2}$ in Chinese sources and 7/4 for $\sqrt{3}$ in Indian ones. The question of whether some such numbers might inherently be inexpressible except in approximate form—what we would now call the existence of irrational numbers—was addressed (or not) in various ways, some of which are discussed in the following section.

Many practical computations, especially of astronomical quantities involving irregular planetary motions, were not amenable to exact solutions in medieval mathematics. They were performed instead using diverse approaches, from simple linear approximations to interpolation schemes using algebraic or trigonometric expressions to elaborate iterative algorithms. Not being easily analyzable by mathematical demonstration, or easily fitted into mathematical classification schemes, such methods were seldom discussed theoretically but frequently invoked in practice.

The ancient technique of "false position" as a sort of proto-algebra approach to finding unknown quantities was well known in several ancient cultures. If the relation of the unknown to the known quantities in such a problem was non-linear, the answer would be approximate. The process of iterating such an approximation to produce successively more accurate answers is attested in a few surviving applications from antiquity, such as "Heron's method" for extracting square roots of non-square numbers. Versions of this method, which involves repeatedly dividing the given non-square number by a guessed value of its square root and then taking the mean of the quotient and the guessed value to produce the next value, were used by ancient Babylonian, Chinese, and Indian mathematicians. Its westward dissemination left traces in the works of, e.g., al-Khwārizmī and Regiomontanus.

Besides the above-mentioned standard place-value root-extraction algorithms applied to decimal or sexagesimal numbers, related methods for root-finding included the Chinese *zengcheng kaifangfa* (Method of [successive] additions [of] multiplicative [powers] to extract [the root]), similar to the Ruffini-Horner method for solving polynomial equations. This was already used by Jia Xian in the eleventh century, for which purpose he devised his "Pascal" triangle (see Figure 3.4). Similar techniques were employed by al-Samaw'al of Baghdad in the twelfth century. The sexagesimal iteration for determining the sine of one degree developed by the fifteenth-century Persian mathematician al-Kāshī, with additions by al-Kāshī's Timurid patron Ulugh Beg, made its way into Indian trigonometry in the seventeenth and eighteenth centuries, where it was further elaborated in the works of Nityānanda of Delhi and Jagannātha of Jaipur.

Fixed-point iteration in general, especially to approximate a wide variety of trigonometric formulas, was a staple of medieval Indian mathematical astronomy. Such approximations were called *asakṛt* ("not just once, repeatedly") in Sanskrit, a name that was also applied to iterated double-false-position and related algorithms. There is some evidence of empirical attempts to analyze and adjust the convergence speed of iterations, particularly in the work of Parameśvara in the Indian Kerala school of the fifteenth century. Some iterative algorithms made their way into Islamic mathematical-astronomy works. Among non-iterative techniques, the most notable for mathematical sophistication were the endless modifications of quadratic approximations to trigonometric formulas used in Indian astronomy from the seventh century onward.

Linear interpolation was the computational mainstay of compilers of numerical tables, particularly calendric and astronomical ones. Table compilation in Chinese and Indian traditions often invoked second- and even third-order interpolation schemes as well. Polynomial interpolation was especially crucial in Chinese astronomical systems, which relied heavily on fitting models of celestial motions between empirically observed positions. Second-order interpolation procedures, some of them probably derived from Indian sources, also appeared in Islamic mathematical astronomy: e.g., the quadratic interpolation formula employed by Brahmagupta in his seventh-century Sanskrit treatise and by al-Bīrūnī, who as we have seen was familiar with Brahmagupta's work, in the eleventh century.

INFINITELY LARGE AND SMALL QUANTITIES

Aristotelian notions of mathematical infinity encompassed infinity of (arithmetic) multitude, infinity of (geometric) magnitude, and infinite subdivisibility of magnitude. Related concepts played a role in all medieval mathematical traditions, whether independently of or influenced by Hellenistic philosophy. Infinitely repeatable or subdivisible processes supplied many medieval mathematicians with concepts of indefinitely reducible quantities reminiscent of early modern infinitesimals, especially in deriving geometric formulas. Concepts of enumerability and infinity were also sometimes applied to extrapolating beyond known large numbers, and the problems of how to express numerical quantity at such scales.

The ancient Chinese Daoist philosophical text *Zhuangzi* propounds some paradoxes of infinity; for example, "if a stick a foot long is cut in half every day, it will not be exhausted in ten thousand generations." Liu Hui's third-century commentary on the *Nine Chapters* employs a comparable form of infinite, or at least indefinite, subdivision in a discussion of approximating circles by polygons, similar to the classical method of exhaustion:

> The larger the number of sides, the smaller the difference between the area of the circle [and that of its inscribed polygons]. Dividing again and again until it cannot be divided further yields a regular polygon coinciding with the circle, with no portion whatever left out. (Shen 1999: 88)

Similar notions were invoked in India, e.g., in Bhāskara's twelfth-century computations of quantities such as the surface area of a sphere (imagined as composed of small pyramids with their vertices at the sphere's center and their bases forming its surface) and instantaneous velocities of planets. Such rationales achieved perhaps their most sophisticated and powerful form in the derivation of infinite series expressions for trigonometric quantities in the work of Mādhava's school in Kerala (southwest India) in the fourteenth–seventeenth centuries: e.g., the Mādhava-Leibniz series for the circumference of a circle, expounded by means of arbitrarily many arbitrarily small right triangles approximating small arcs of the circumference. In Greco-Islamic mathematics, quasi-Archimedean exhaustion arguments were central to results such as the geometric derivation of the volume of a paraboloid by Ibn al-Haytham around 1000.

Medieval European Aristotelianism retained its founder's anti-atomistic suspicions of the concept of indivisibility. For most of those philosophers, therefore, physical and mathematical continua had to be infinitely divisible. But the idea of actually existing infinity, although rejected by Aristotle, was doctrinally established in the limitlessness of the Deity, and some scholastic philosophers suggested that it might also be embodied in contexts such as the number of points on a continuous line. The fourteenth-century Merton College philosopher Thomas Bradwardine summarized these arguments in part as follows:

> One must know that the old and modern philosophers have five famous opinions about the composition of the continuum. Some of them, such as Aristotle, Averroes, Algazel [eleventh-century philosopher al-Ghazālī] and most of the moderns, argue that the continuum is not composed of atoms, but of parts that can be divided without end . . . Others say that it consists of points, and this [assumption is divided] into two parts: Pythagoras (the father of this position), Plato, and our contemporary Walter [Chatton] assume that it is composed of a finite number of indivisibles, but others say that [it is composed] of an infinite number . . . (Katz 2017)

Bradwardine argues against the hypothesis of a finite number of indivisibles, which would imply, for example, that half the circumference of a circle equals its diameter, because only one perpendicular can be drawn to the circumference from any point on the diameter. He likewise rejects the supposition that the continuum is composed of infinitely many indivisibles, because then parallelograms with equal bases and equal heights could have unequal areas depending on the lengths of the infinitely many parallel lines composing those areas. He concludes:

> No continuum is made up of atoms. From here follows and elicits: Every continuum is composed of an infinite number of continua of the same species as it, [. . .] that is, every line is composed of an infinite number of lines, every

surface composed of an infinite number of surfaces, and so on concerning other continua. (Katz 2017)

The hypothesis of an infinite quantity of infinitesimal magnitudes was perhaps most famously deployed in the quadrature of the circle as an infini-lateral polygon proposed by Nicholas of Cusa in the early fifteenth century.

Actual mathematical infinity, though rejected by Aristotelians such as Ibn Sīnā (Avicenna) and Bradwardine, made some interesting appearances in other mathematical contexts. Its relation to division by zero was explained by Bhāskara II as follows:

> In this quantity also, which has zero as its divisor, there is no change even when many [quantities] have entered into it or come out [of it] just as at the time of destruction and of creation, when throngs of creatures enter into and come out of [him, there is no change] in the infinite and unchanging one [i.e., Lord Viṣṇu]. (Plofker 2007: 471)

FOUNDATIONS AND METHODOLOGIES

How is mathematics made? It used to be commonly assumed by modern mathematicians and historians of mathematics that the Euclidean structure of axioms, propositions, and rigorous demonstrations is necessary for the attainment of "genuine" mathematical truth. Dirk Struik's 1967 dicta, "Nowhere in all ancient Oriental mathematics do we find any attempt at what we call a demonstration . . . Oriental mathematics never seems to have been emancipated from the millennial influence of the problems in technology and administration, for the use of which it had been invented," (Struik 1967: 30) reflect the extent to which Hellenistic ontological and epistemological frameworks are the water that modern mathematicians swim in: so ubiquitous and familiar as to be invisible. In this section we also consider some of the many other ways that mathematical knowledge was organized and validated in the medieval period.

Explicitly axiomatic-deductive proof structure was not employed in Chinese mathematics until the introduction of Western mathematical texts by Jesuits in the seventeenth century. Neither was the type of counterfactual argument commonly known as "proof by contradiction." Instead, Chinese mathematicians centered the notion of *bilei*, generally translated "analogy" or "categories," in structuring and logically justifying results. Different methods and problems were broken down into shared components which could be handled in similar ways to provide solutions. This was a more sophisticated approach than mere naive *argumentum ad exemplum*. Geometric shapes, for example, might be dissected into standardized sub-units in order to apply validated computational formulae to them.

This technique of analogous reasoning was described as early as the work of Liu Hui:

> By properly arranging concepts and propositions through analogous and deductive analysis, they could be put in their proper place. Therefore those branches which grow diversified but share the main stem of a tree are comprehended to be from one origin. Furthermore to analyze a theory by proposition, and to illustrate a structure with geometrical figures, will thereby make the whole picture of the theory or structure be understood through some simple principles . . .
>
> The proposition comes out from direct causes; develops with general laws or theory; and is manipulated by applying to the same class of [things] . . . (Wann-Sheng Horng, trans., quoted from Dauben 1998: 1350–1)

The thirteenth-century mathematician Yang Hui reorganized and expanded areas of mathematical knowledge based on these analogical categories, inspiring a new branch of mathematics in the subsequent centuries.

Various visual and manipulative tools supported the comprehension of verbal and numerical reasoning: for instance, the previously mentioned use of the colors red and black for counting rods to distinguish between positive and negative quantities. Color markings were also used in geometric representations known as *tu*, which may have started out as separate dissected paper shapes, alluded to in ancient texts such as Liu Hui's. Although often lost from the earliest Chinese mathematical classics, such figures in many cases may be reconstructed. In later block-printed texts beginning with the Song dynasty, they had evolved into two-dimensional geometric figures drawn within the texts themselves. Physical models of four basic geometric solids with known volume formulas—the cube, right prism, right pyramid, and *bienao*, the four-sided figure that remains after cutting a right pyramid from a right prism—served to justify visually computations of the volumes of three-dimensional solid figures composed from them.

Visual aids in these later texts could also include tables representing data of various kinds, and drawings of counting-rod layouts (likewise called *tu*) to illustrate computational algorithms: i.e., how the numbers or counting rods would have been set out on the counting board, and how they would have appeared in different configurations before and during computation. Ancient references to three-dimensional manipulatives (*qi* "blocks") seem to have been discontinued in the late medieval works.

Indian strategies for the validation of mathematical knowledge were also very different from Euclidean proof, but likewise involved a far more critical and insightful approach than has been recognized in many histories of "oriental mathematics." The Sanskrit verse format of medieval Indian texts on mathematics (and many other subjects), like the *sūtra* format of their ancient counterparts, conduced to brief summaries of formulas and algorithms. Rationales (Sanskrit *yukti, upapatti*) explaining and justifying them might be left to prose commentaries or to oral exposition, or omitted if the textual content seemed sufficiently self-explanatory; they focused on the experience of comprehension and persuasion rather than on a formal hierarchy of propositions.

The significance of such demonstrative reasoning is illustrated by the following passages in the twelfth-century mathematical-astronomy work *Siddhāntaśiromaṇi* of Bhāskara. He remarks in his auto-commentary that "it is said that in the realm of *gaṇita* a precept [is acceptable as] valid knowledge (*pramāṇa*) only [by virtue of] having a demonstration (*upapatti*)." In a later verse he reiterates the point by contrasting *gaṇita* with *gola* "sphere," i.e., the accompanying spherical astronomical models, which he likens to the round *amla* or Indian gooseberry:

> A mathematician (*gaṇaka*) [knowing only] the calculation (*gaṇita*) of the planets [stated] here [in the chapters on] mean motions and so forth, without the demonstration (*upapatti*) of that, will not attain greatness in the assemblies of the eminent, [and] will himself not be free from doubt. In the sphere, that clear [demonstration] is perceived directly like an *amla* [berry] in the hand. Therefore I am undertaking the subject of the sphere as a means to understanding demonstrations. (Plofker 2009: 197)

This sentiment is closely echoed in the context of non-astronomical mathematics by Gaṇeśa, a sixteenth-century commentator on Bhāskara's arithmetic work, *Līlāvatī*:

> In *vyakta* [mathematics of knowns] or [that] called *avyakta* [mathematics of unknowns], the one [who knows] completely that which is stated, [but] without demonstration, [is] not free of doubt concerning [its] truth and will not attain greatness in the assembly of good mathematicians. (Āpaṭe 1937: vol. 1, 1)

The concept of *pramāṇa*, i.e., a valid means of attaining exact knowledge, is central to Indian philosophy. In the category of *pramāṇa* are included direct sense perception, inference, verbal authority, and other forms of inference or analogy, depending on the philosophical school. (Proof by contradiction was generally not recognized among these.) Physical objects and diagrams could help guide the student to correct perception, like the abovementioned *amla* berry whose round shape and longitudinal stripes were sometimes invoked to visualize the globe. But detailed rigorous expositions and proliferation of deduced rules seem to have been viewed more as a concession to less gifted minds than as a required confirmation of truth, which the intelligent could discern even in its most abstract form. As Bhāskara modestly remarks at the end of his Rule of Three exposition quoted near the start of this chapter:

> What has been composed by the sages through the multifarious methods and operations such as miscellaneous rules, etc. teaching its easy variations, is simply with the object of increasing the comprehension of duller intellects like ourselves. (Datta and Singh 1962: 209)

Indeed, more subtle approaches to rational demonstration may have been built into the structure of Sanskrit texts: e.g., the "apodictic discourse" model that has been proposed for some passages in Brahmagupta's treatise, where sequences of apparently stand-alone rules in concise verses form an implied progressive exposition. However, it is also possible that versified computational formulas occasionally became decoupled from the details of accompanying demonstrative rationales. Certain cyclicity conditions on some of Brahmagupta's rules about quadrilaterals, for example, appear to have been unknown to, or rejected by, some later authors.

A sort of "Euclidean-supremacist" distrust of such flexibly structured approaches to mathematical demonstration pre-dated by many centuries the criticisms of moderns such as Struik. The renowned al-Bīrūnī, for example, declared in his *India*:

> I began to show [Indian scholars] the elements on which this science rests, to point out to them some rules of logical deduction and the scientific methods of all mathematics, and then they flocked together round me from all parts, wondering, and most eager to learn from me . . . I can only compare their mathematical and astronomical literature, as far as I know it, to a mixture of pearl shells and sour dates, or of pearls and dung, or of costly crystals and common pebbles. Both kinds of things are equal in their eyes, since they cannot raise themselves to the methods of a strictly scientific deduction. (Sachau 1992: 23, 25)

Well before al-Bīrūnī's time, the Indian works on mathematical sciences that were studied in the early Islamic world had mostly been succeeded by Hellenistic sources. Arabic versions of Greek logic and epistemology were absorbed along with the technical writings of mathematicians such as Euclid and Ptolemy. In particular, the Aristotelian views of Ibn Sīnā in the eleventh century largely displaced the tendency toward Pythagoreanism and Platonism in Islamic philosophy of mathematics.

The goal of reliably systematizing the discovery and proof of mathematical results was pursued in treatises on analysis and synthesis by, among others, Ibrāhīm ibn Sinān, al-Sijzī, and Ibn al-Haytham. A valid Euclidean proof remained the epistemological gold standard, as proclaimed by the fourteenth-century Maghrebi scholar Ibn Khaldūn:

> Geometry enlightens the intellect and sets one's mind right. All its proofs are very clear and orderly. It is hardly possible for errors to enter into geometric reasoning, because it is well arranged and orderly. Thus the mind that constantly applies itself to geometry is unlikely to fall into error. In this convenient way, the person who knows geometry acquires intelligence. (Ibn Khaldūn 2015: 378)

However, not all Islamic mathematical works hewed to classical ideals of geometric demonstration. For instance, treatises by several authors including al-Karajī and Ibn al-Bannā' proved algebra rules solely by arithmetical arguments, sometimes remarking that geometric ones might be too unfamiliar or difficult for the reader. On the other hand, some classical practices might be rejected on mathematical or physical grounds for not being rigorous enough. The tenth-century geometer Abū Sa'īd al-Sijzī, for example, objected to "neusis" or "verging" constructions in geometry, which permitted the sliding of points or lengths in ways unsupported by Euclid's postulates. And Islamicate astronomers frequently complained about violations of Aristotelian physical principles in Ptolemaic orbital models.

The prospect of innovation in methods of proof was not without appeal in late medieval European mathematics either. Early forms of mathematical induction techniques employed by Abū Kāmil, al-Karajī, and others preceded, and perhaps influenced, a version developed by Levi ben Gerson. We have mentioned in a previous section the "art" of Ramon Llull for systematizing logical discourse, although it was intended for theological arguments rather than specifically mathematical ones.

Pre-Renaissance Latin mathematics consciously derived its philosophical foundations both from ancient authorities—above all, "The Philosopher" himself, Aristotle—and from their interpreters and successors in the Islamic world, particularly Ibn Sīnā and Ibn Rushd (Averroes). These views were not without their critics, among them Roger Bacon in the thirteenth century, who argued for a revised taxonomy of mathematical knowledge. The first part of the subject should be what Bacon called "common mathematics" expounding the concepts shared by all mathematical fields; the second should comprise the quadrivium subjects in both "speculative" and practical forms. Mathematics for Bacon was the "gate and key" to all other quantitative sciences, foreshadowing to some extent the concept of mathematics as the "language of nature" that dominated the approaching Scientific Revolution.

CHAPTER FOUR

Mathematics and Worldviews

Medieval Views of the World—In the Heavens and on Earth

JOSEPH W. DAUBEN

> The square earth is my chariot, and
> the round heaven is my canopy—Song Yu, fl. fourth century BCE.
> (Cullen 2017: 203)

Nothing distinguishes the differences in worldviews between East and West more dramatically than does the Chinese belief that the earth was square and basically flat, compared with the Greeks' realization that the Earth must be spherical. From ancient times in China, and throughout the medieval period—and well into the Ming and Qing dynasties—the view expressed by Song Yu prevailed, that the *yin/yang* dichotomy was reflected in the heaven/earth dichotomy of a circular spherical heaven and a square flat earth. In the medieval Latin West the Ptolemaic view was generally accepted—a world divided into a terrestrial sub-lunar realm (comprised of four elements subject all the while to generation and corruption), whereas above the moon, the heavens were belived to be comprised of crystalline spheres made of a presumably incorruptible element, aether (whose nature was to move always and forever with constant, continuous, uniform motion in combinations of perfect circles). In sharp contrast to this was the view of the cosmos that was generally accepted in China throughout the medieval period. Heaven and earth were believed to be intimately connected through the persona of the ruling emperor, and was far from being regulated by any strictly mechanical or clockwork vision of the universe.

WORLDVIEWS IN MEDIEVAL CHINA

The earliest model of the heavens and earth to be given any mathematical precision is described in considerable detail in the most ancient of early Chinese works devoted to astronomy, the *Mathematical Classic of the Zhou Gnomon*. This work has survived from the Western (Early) Han dynasty, and presents the details of the *gaitian* (canopy-heaven) cosmos. This pictures the heavens as a circular disc 80,000 *li* above a square, flat earth below. The *gai* of *gaitian* refers to the umbrella-like cover of ancient Chinese chariots, the cover suspended on a pole above the square frame of the chariot itself. Such chariots have been recovered from early Qin and Han dynasty tombs, and what have been called "cosmic models" related to this theory have also been excavated from ancient burial sites. Manufactured from bronze or painted on lacquered wood, one such object was recently found in the tomb of Xiahou Ying, the Marquis of

FIGURE 4.1. Bronze chariot and horses, Museum of the Terracotta Army, Xi'an, Shaanxi Province, China. Photo by Mel Longhurst / VW Pics / Universal Images Group / Getty Images.

Ruyin, and is comprised of two lacquered wooden plates, the lower plate being flat and square representing the earth, the upper plate a disk representing the heavens and meant to turn on a pin mounted at the center of the bottom plate.

Based upon the assumption that the earth was flat, by comparing the lengths of shadows cast by two gnomons a known distance apart, it was determined in the *Mathematical Classic of the Zhou Gnomon* that the sun and therefore the heavens must be some 80,000 *li* above the earth (about 25,000 miles). This may be compared with the estimate made by Ptolemy that the sphere of fixed stars was some 20,000 earth radii away (nearly 60,000,000 miles). The difference in worldviews measured in terms of just these dimensions alone is dramatic.

To explain why the sun and other heavenly bodies, thought to be imbedded in the umbrella 80,000 *li* above the earth, were not always visible, the *Zhou Gnomon* held that beyond a certain distance, 167,000 *li* to be exact, the sun, moon, and planets were beyond the range of human sight. In Christopher Cullen's words, their risings and settings were taken to be illusory (Cullen 2017: 212).

Aside from the ad hoc nature of this explanation, the *gaitian* theory was subject to other more serious objections, voiced as early as the Eastern (Later) Han dynasty administrative official Huan Tan, who not only objected that the theory failed to account for the equinoxes as observed by anyone not directly under the point on which the heavens were assumed to revolve, but it was also contradicted by the observed ascent of the sun from early morning to high noon. As Cullen summarizes Huan Tan's argument:

> If the sun was just moving horizontally as the disc of heaven rotated [according to the *gaitian* theory], why should that happen? Was not this observation in better accord with the *hun tian* notion that the heavens were a sphere – so that the sun's markedly increasing altitude would be a simple consequence of it rising up over the horizon with the turning heavenly sphere? (Cullen 2017: 221)

Among those who advanced this rival *huntian* (celestial sphere/egg-like) cosmos was Zhang Heng. In this model of the universe, the earth was taken to be surrounded by a curved sphere rotating over it. Zhang Heng is reported to have described it all as follows:

> Heaven is like a hen's egg, and the earth is like the yolk inside. It is alone within; heaven is the larger, and earth the smaller. There is water inside and outside heaven; Heaven and earth both stand by riding on *qi*, and move

borne up by water. (*Sui shu* 19, 509; quoted from Cullen 2017: 280)

A third model of the cosmos, one more abstract and later than the others, is the *xuanye* (infinite empty space/firmament) cosmos. According to this theory, the heavens are taken to be infinite in extent, and the celestial bodies were regarded as floating on and moved by *qi*, something akin to pneuma or a very subtle form of matter, variously translated as vapor, air, wind, exhalations, vital energy, etc. A general description of this model provided by the Eastern Han cosmologist Ji Meng (Needham: Chhi Mêng) appears in the astronomical chapters of the *Jin shu*—the *Book of Jin*, a history of the Jin dynasty (265–420 CE). The early fourth-century master Ge Hong (Needham: Ko Hung) wrote of the *xuanye* theory as follows:

> They said that the heavens were empty and void of substance (*wu chih*). When we look up at it we can see that it is immensely high and far away, having no bounds (*wu chi*). . . . The sun, the moon, and the company of the

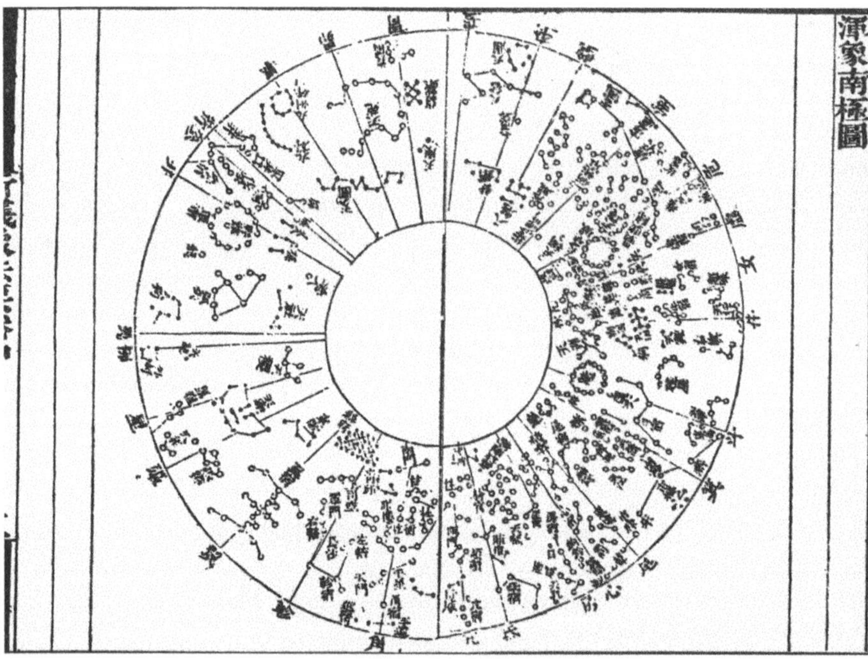

FIGURE 4.2. Star-map from the south polar projection of a celestial globe, an instrument featured as part of Su Song's celebrated clocktower (see Chapter 5) and published in his book about this device, *Essentials of a New Method for Mechanizing the Rotation of an Armillary Sphere and a Celestial Globe* (1092) (Needham 1959: 277). Courtesy of the Needham Research Institute, Cambridge, UK.

> stars float (freely) in the empty space (*fou khung chung*), moving or standing still. All are condensed vapour (*chieh chi chhi*). Thus the seven luminaries sometimes appear and sometimes disappear, sometimes move forward and sometimes retrograde, seeming to follow each a different series of regularities; their advances and recessions are not the same. It is because they are not rooted (to any basis) or tied together that their movements can vary so much. . . . Their speed depends on their individual natures, which shows that they are not attached to anything, for if they were fastened to the body of heaven, this could not be so. (Needham 1959: 219)

Needham summarizes the divide between Greek and Chinese thought about the heavens this way:

> In a word, the Chinese astronomers were practically free from the cramping orthodoxy of Hellenistic and medieval Europe which supposed that the heavenly bodies were fixed to a series of concentric material spheres with the earth as centre. These spheres were essentially nothing but an unjustified concretization of spherical geometry, and it is rather paradoxical that the Chinese, whose views are so often accused of being excessively materialistic and concrete, should have remained comparatively free from them. They may have had no deductive geometry, but they had no crystalline spheres either. (Needham 1959: 223)

WORLDVIEWS IN MEDIEVAL INDIA

Chinese religious cosmographies, to which Needham devotes an entire section in the part of volume 3 concerning astronomy in his monumental series *Science and Civilisation in China* (Needham 1959: 565–8), shared with their Indian counterparts elements of numerical mysticism. Chiefly through Buddhist doctrine, these Indic models had a conspicuous influence later in medieval China. The linchpin of their structure, literally and figuratively, was the immense mountain called Mt. Meru at the center of the world, which lends its name to the Meru cosmos, or *cakravāla* ("wheel-system") model of the universe. The *cakravāla* model appears throughout Buddhist thought, practice, and material culture, in different versions depending upon place and time that have influenced aspects of daily life such as architecture and ritual.

In the most ancient of the surviving Vedic Sanskrit works, the *Ṛgveda* hymns dating back to at least the second millennium BCE, various accounts of the creation of the world suggest other models. The Vedic cosmos has three worlds or *lokas*: the world of earth and the world of heaven are separated by the intermediate region or atmosphere containing air and light. Some hymns invoke

FIGURE 4.3. Mandala, or cosmic diagram, illustrating Indo-Himalayan imagery centered on the mythological Mount Meru, represented by an inverted pyramid topped by a lotus, Buddhist symbol of purity. Traditional Chinese images for the sun (three-legged bird) and moon (rabbit) appear at the mountain's base. Courtesy of the Metropolitan Museum of Art, New York, NY. Photo by History / Universal Images Group / Getty Images.

cosmic metaphors such as the growth of the world from a golden embryo or germ, or from the dismemberment of the primeval Cosmic Man; weaving shuttles constructing the world of the sacrifice; a turning wheel representing the year with 360 paired days and nights standing on its twelve spokes. Or the world may be represented as a building made of wood constructed by the gods, with heaven and earth supported by posts (North 2008: 171). The year, the synodic month, and the day were fundamental units of a Vedic ritual calendar system whose cycles were marked by full and new moons, solstices and equinoxes. Larger time cycles and the idea of periodic creation and destruction of the universe are not systematically described in these earliest texts, although the concept of a *yuga* or multi-year period is attested.

At some point in the ancient development of the sacred compendia of historical legend and myth known as the Purāṇas ("ancient ones"), which had attained their canonical form by the first few centuries of the Common Era, Vedic and post-Vedic ideas of number and the physical world crystallized into the Meru cosmos. Its universe is an egg-shaped form enclosing the flat disk of the earth with the great mountain Meru in its center, surrounded by the primal elements. All the dimensions involved are immense: for example, the diameter of the earth's disk is said to extend for 500 million of the units called *yojanas*, which would be approximately on the order of 5 billion kilometers. Mt. Meru reaches to the pole-star in the heavens, and the other stars and planets wheel around it, appearing to rise or set as they are revealed or hidden by its massive form.

In this cosmology the sun is taken to be nearest the earth, the moon behind it, then the *nakṣatras* (the twenty-eight lunar mansion asterisms) followed by the five planets in the sequence Mercury, Venus, Mars, Jupiter, and Saturn. Above all these and above Mt. Meru are the "Seven Sages" (or Big Dipper), and beyond these the higher layers of the heavens. The flat earth is shown as a series of concentric rings or squares, usually seven in number, of land separated by spans of water and other liquids between them. As if to explain why there are so many different versions of this view of the cosmos, the *Vāyu Purāṇa* points out that no one really knows the exact configuration of the world: "Only Brahmā knows it in full" (Greenwood 2015: 32).

Temporal aspects of the Meru cosmos reflect the same sort of numerical vastness as its spatial aspects. Both the universe and its human inhabitants are subject to a constant repetition of death and rebirth. The largest cycle or period between creation and destruction of the universe, the *kalpa*, is said in both the *Viṣṇu Purāṇa* and the *Bhāgavata Purāṇa*, to last for 4.32 billion years, or 1,000 *mahayugas*. Each *mahayuga* is composed of four smaller *yugas* with successively decreasing durations in a 4:3:2:1 ratio, characterized by continued deterioration from an original idyllic state which culminates in a 432,000-year Kaliyuga plagued with wickedness and suffering. The current Kaliyuga, which in most forms of this Hindu cosmology is placed around the middle of the current *kalpa*, is held to have begun about 5,000 years ago, at the time of the great war recounted in the *Mahābhārata* epic.

The first extant detailed quantitative models of calendric cycles in Sanskrit texts date probably from around the middle of the first millennium BCE (although various earlier dates have also been proposed on the basis of differing astronomical interpretations of the calendar schemes described in the text). Their references to instruments such as gnomons and water clocks may indicate influence from Mesopotamian astronomy transmitted via the Achemenid Empire.

FIGURE 4.4. Cakravāla mural, depicting the Buddhist universe with Mt. Meru surrounded by seven rings of golden mountains and blue oceans, Trongsa Dzong, Trongsa, Bhutan. Photo by Richard l'Anson/Getty Images.

After the incursions of Alexander, Indo-Greek settlements in the northwest of the subcontinent also introduced Hellenistic systems of spherical astronomy and horoscopic astrology. During the first few centuries of the Common Era, the Vedic ritual calendar, Purāṇic long cosmic periods, Hellenistic astronomical coordinate systems and orbital models, as well as Greco-Roman astrology including time units such as the seven-day week, were synthesized into classical Indian mathematical astronomy/astrology or *jyotiṣa*. As described in Chapter 3, this transformation included the creation of trigonometry of sines for solving plane right triangles within the sphere.

The canonical text genre expounding these mathematical models was the Sanskrit astronomical *siddhānta* or treatise, which assumed a great conjunction of all celestial bodies at the beginning of the current *kalpa*, and integer numbers of orbital revolutions about the earth for each of them in the course of a *kalpa*. Spherical geometry and trigonometric orbital models, along with the stated *kalpa*-parameters and a knowledge of one's own temporal location within the *kalpa*, enabled the user to compute predicted positions and appearances of celestial bodies at any time. Their configurations could then be interpreted astrologically to understand what they portended.

The treatise *Āryabhaṭīya* of the early sixth-century mathematician-astronomer Āryabhaṭa preserves the earliest extant example of such a complete astronomical system or *siddhānta*. The last 50 of the *Āryabhaṭīya*'s 118 verses are devoted to *gola* (the sphere), primarily the nature of a spherical earth and the problem of eclipses. The worldview that Āryabhaṭa presented was especially remarkable for its suggestion that the earth rotates on its axis. His explanation for how this could happen without an observer's awareness that the earth was moving prefigures Galileo's famous explanation of relative motion:

> In the same way that someone in a boat going forward sees an unmoving [object] going backward, so [someone] on the equator sees the unmoving stars going uniformly westward. The cause of rising and setting [is that] the sphere of the stars together with the planets [apparently?] turns due west at the equator, constantly pushed by the cosmic wind. (*Āryabhaṭīya*, quoted from Plofker 2009: 111)

Āryabhaṭa's rotating-earth hypothesis was not embraced by most later Indian astronomers, who favored geostationary models on the basis of intuitive physical arguments. The 629 *Āryabhaṭīyabhāṣya* of Bhāskara I, a major commentary on the *Āryabhaṭīya*, supported Āryabhaṭa's view of a spherical cosmology:

> "In the center of the heavens": The earth is not at all above [the center], and not below, hence it is not falling . . . Now others think [that] the earth is supported by [the cosmic serpent] Śeṣa or [something] else: that is not rational . . . Now if they [i.e., Śeṣa etc. can] stay fixed by their own power, why cannot this power be assumed for the earth? (*Āryabhaṭīyabhāṣya*, quoted from Plofker 2009: 112)

Later adjustments to the siddhāntic worldview included a modified geocentric system, with the interior planets Mercury and Venus revolving about the sun rather than the earth, in the early sixteenth-century astronomy of Nīlakaṇṭha of Kerala. But the fundamental geocentrism and sphericity of the universe and its component bodies remained a chief principle of mathematical astronomy in Sanskrit *jyotiṣa* throughout the medieval and early modern periods, coexisting more or less uneasily with the traditional Meru cosmos of sacred works such as the Purāṇas.

ISLAMIC MODIFICATIONS OF PTOLEMAIC COSMOLOGY

Sasanian and Indian forms of spherical astronomy appear to have provided the first mathematical models of Islamic astronomy, which soon afterwards began to acquire Hellenistic versions directly from translated Greek sources. Sasanian and Indian forms of spherical astronomy appear to have provided the first mathematical models of Islamic astronomy, which soon afterwards began to acquire Hellenistic versions directly from translated Greek sources.

Modifications to Ptolemaic astronomy in the Islamic world, along with better tables, improved instruments, and more accurate observations, all served to expose weaknesses in both Aristotelian and Ptolemaic worldviews. In Baghdad, under the 'Abbāsid caliphates of al-Manṣūr and al-Ma'mūn, astronomy was advanced due to specific needs of Islam—determining times for prayer and directions to Mecca, moments of sunrise and sunset fixing the limits for fasting during Ramadan, and first appearances of the moon signaling the start of a new month (see further details in Chapter 2).

Important improvements to Ptolemaic models were made, for example, by Aḥmad al-Farghānī (Alfraganus), an astronomer/astrologer at the court of al-Ma'mūn. Al-Farghānī is known for having written a book on the construction of sundials, a summary of the *Almagest*, *Kitāb al-Fuṣūl, ikhtiyār al-Majisṭī* (The Book of Chapters, a Summary of the *Almagest*), several treatises on the astrolabe, and a commentary on the astronomical tables of al-Khwārizmī. His *Kitāb fī Jawāmī' 'ilm al-Nujūm* (Elements of Astronomy on the Celestial Motions, *c.* 833), was his most influential work. Latin translations were later made by John of Seville and another by Gerard of Cremona. The first printed edition of the translation by John of Seville appeared in 1493 and was reprinted several times thereafter.

Al-Farghānī worked with al-Khwārizmī and other astronomers in the *Bayt al-Ḥikmah* (The House of Wisdom) in Baghdad. He took part in measuring the length of a degree of the meridian in the Sanjar desert of Iraq, and as a result, in his *Elements* he was able to calculate the circumference and the diameter of the earth as approximately 20,400 miles and 6,500 miles, respectively. He also surpassed Ptolemy in estimating the distances of the planets from the earth, in addition to the distances of the sun and moon that Ptolemy had given.

Abū Rayḥān Muḥammad ibn Aḥmad al-Bīrūnī was a Persian polymath who travelled as far as India, and was among those whose worldviews did not exclude the possibility of a moving earth. Although al-Bīrūnī admitted that he could neither prove nor refute the motion of the earth, he did admit that there was nothing contradictory in assuming its rotation. Substantially influenced by astronomy from India, he devoted to the subject large portions of his book *Taḥqīq mā li-l-hind* (Verification of What Pertains to India), drawing largely on the works of Āryabhaṭa and Brahmagupta. In it, al-Bīrūnī summarized the situation as he saw it as follows:

> Brahmagupta says in another place of the same book: "The followers of Āryabhaṭa maintain that the earth is moving and heaven resting. People have tried to refute them by saying that, if such were the case, stones and trees

> would fall from the earth." But Brahmagupta does not agree with them, and says that that would not necessarily follow . . . Besides, the rotation of the earth does in no way impair the value of astronomy, as all appearances of an astronomic character can quite as well be explained according to this theory as to the other. There are, however, other reasons which make it impossible. (Sachau 1910: vol. 1, 276–7)

Ḥasan ibn al-Haytham (Alhazen), one of the most prolific of medieval Muslim mathematicians, although best known for his work on optics, was also referred to as a "second Ptolemy" and, by the thirteenth-century Archbishop of Canterbury, John Peckham, as "the physicist." In a synopsis of the Ptolemaic theory, *Hay'at al-'ālam* (On the Configuration of the World), al-Haytham held that the earth was a sphere, fixed at the center of the universe and motionless.

However, in his *Al-Shukūk 'alā Baṭlamyūs* (Doubts Concerning Ptolemy), al-Haytham noted various contradictions and objections he had to Ptolemy's *Almagest* and *Planetary Hypotheses*, as well as the *Optics*. In an anticipation of Copernicus, he was especially critical of Ptolemy's introduction of the equant, the imaginary point with respect to which uniform motion could be assumed in order to account for the non-uniform motions of the planets. Al-Haytham held this to be a basic violation of the most fundamental axiom of ancient astronomy, that the heavens must move continuously in uniform perfect circles. He also rejected the use of imaginary points, lines, or circles in trying to account for any physical motions of the heavens, including the epicycles moving on deferent circles that were essential for Ptolemy's geometric model of planetary motions.

As for Ibn al-Haytham's general worldview, he maintained a healthy skepticism and urged that one keep an open mind, doubting everything in the pursuit of certainty. Although valuing Ptolemy, he realized that:

> Truth is sought for itself, but the truths are immersed in uncertainties and are not immune from error. Therefore, the seeker after the truth is not one who studies the writings of the ancients and, following his natural disposition, puts his trust in them, but rather the one who suspects his faith in them and questions what he gathers from them, the one who submits to argument and demonstration, and not to the sayings of a human being whose nature is fraught with all kinds of imperfection and deficiency. Thus the duty of the man who investigates the writings of scientists, if learning the truth is his goal, is to make himself an enemy of all that he reads, and, applying his mind to the core and margins of its content, attack it from every side. He should also suspect himself as he performs his critical examination of it, so that he may avoid falling into either prejudice or leniency. (Sabra 2003)

A partial version of another of al-Haytham's astronomical works has survived, *The Model of the Motions of Each of the Seven Planets.* Rejecting any highly abstract mathematical vision of the *Almagest*, al-Haytham approached his theory of planetary motions in terms of spherical geometry. His model was geocentric and made use of epicycles, but eliminated the equant. Rather than explain the causes of planetary motions, his model was purely geometric and sought to account for the appearances of observed phenomena, thereby "saving the phenomena" but on terms that involved no unwarranted assumptions like the equant.

Muḥammad ibn Muḥammad ibn al-Ḥasan Naṣīr al-Dīn al-Ṭūsī, one of the greatest medieval Persian thinkers, was the creator of trigonometry as a mathematical discipline. Raised in a scholarly tradition, in the wake of the Mongol conquest of Persia al-Ṭūsī eventually became the court astrologer under Hūlāgū Khan, whose faith in astrology and belief that many of his military victories were due to the advice of court astrologers led him to support al-Ṭūsī's astronomical work. Al-Ṭūsī had complained that data he was using were more than 250 years old, and so Hūlāgū Khan agreed to construct a new observatory, which al-Ṭūsī was to head. Thus a state-of-the-art astronomical observatory was constructed in Marāgha (Azerbaijan), capital of the Mongol Empire ruled by Hūlāgū. As director of the observatory, al-Ṭūsī also oversaw an affiliated library and school. With the instruments of the new observatory and a retinue of astronomers—some even from as far away as China, like the astronomer Fao Mungji (possibly Fu Muzhai, Fu Manzi, or the astronomer Fu Yanqing)—al-Ṭūsī was able to make very accurate tables of planetary movements comprising the *Zīj-i Īlkhānī*.

Especially noteworthy were al-Ṭūsī's objections to the observational evidence Ptolemy had used to argue that the Earth could not be moving, saying that such evidence was inconclusive and instead, only on principles of natural philosophy, like the immovability of the terrestrial elements, could one conclude that the earth was immovable. Like his sixteenth-century commentator al-Bīrjandī, al-Ṭūsī maintained that the earth's immobility could be demonstrated only by physical principles drawn from natural philosophy. Al-Ṭūsī's criticisms of Ptolemy were similar to arguments later used by Copernicus in 1543 to defend the Earth's rotation (Saliba 1994). Thus al-Ṭūsī was especially influential on his astronomical and cosmological successors thanks to his openness to discussion of the Earth's motion, although he remained committed to a geocentric universe. As was the case with many Islamic scientists, al-Ṭūsī regarded the certitude of the exact mathematical sciences, especially astronomy and pure mathematics, as the best means of appreciating and understanding God's creation.

His greatest invention was the Ṭūsī-couple, a geometric device that used the combined motions of two circles, one within the other, that could simulate back-and-forth linear motion, much like the epicycle-deferent motions of ancient Greek astronomy. The Ṭūsī-couple enabled al-Ṭūsī to avoid the objectionable Ptolemaic equant, except for the case of Mercury, for which a satisfactory non-equant model was later found by Ibn al-Shāṭir and ʿAlī al-Qūshjī. Using the Ṭūsī-couple, al-Ṭūsī was able to reproduce the oscillating back-and-forth retrograde motions of the planets without sacrificing accuracy, but with no need for epicycles and not having to violate the fundamental assumption of ancient astronomy that all heavenly motions must be uniform.

The Ṭūsī-couple first appeared in al-Ṭūsī's Persian treatise, *Ḥall-i mushkilāt-i Muʿīniyya* (Solution of the Difficulties in the *Muʿīniyya*), and was further developed and used again years later in his *al-Tadhkira fī ʿilm al-hay'a* (Memoir on Astronomy). Al-Ṭūsī's devices are of major significance for several reasons. First, they produced models that conformed to preconceived physical and mathematical requirements; they separated the effect of the distance of a planet from its speed (which had been conjoined in the Ptolemaic models). Al-Ṭūsī was thus able, for example, to circumvent Ptolemy's reliance on the circular motion of epicycles to produce a rectilinear, latitudinal effect. Second, al-Ṭūsī's new models greatly encouraged and influenced the work of subsequent Islamic astronomers, such as his student Quṭb al-Dīn al-Shīrāzī and Ibn al-Shāṭir, and possibly even Copernicus as well.

INNOVATIVE WORLDVIEWS IN EUROPE

In his chapter in Volume 1 devoted to the subject of mathematics and worldviews in the ancient world, Geoffrey Lloyd emphasized the special status of Greek mathematics, regarded as the single source of certain knowledge with an axiomatic-deductive methodology that guaranteed the rigor of geometric demonstrations, at least by those working in the Euclidean tradition. Likewise, a syllogistic form of reasoning also gave rigor to certain forms of argumentation in general, as explored by Aristotle in his *Posterior Analytics*. Consequently, mathematics was regarded as a means of accessing profound truths about life and the mysteries of the universe that might assume moral and even religious significance. It was the rigor of the axiomatic-deductive methods of the Greeks that gave those who adopted it a sense of the absolute certainty of mathematics. This extended to the models of the heavens put forward in their most advanced form by Ptolemy in which a clockwork universe should be one that allowed for the most accurate predictions of future events (and similar determination of past events as well). But such certainty was not in general a feature of the worldviews entertained by any other cultures of the medieval period outside of the Greco-Roman and later Islamicate spheres of the Mediterranean and parts of Europe.

Despite the supposed "certainty" of mathematics, this very certainty led to some very unexpected and uncertain consequences. The discovery of incommensurable magnitudes also meant that Greek mathematics would depend upon geometry and marginalize arithmetic, since there were magnitudes that could not be measured by numbers, which the Greeks limited either to whole numbers or ratios of whole numbers, i.e., fractions. By admitting only integers and rationals as numbers, there were no "numbers" corresponding to incommensurable magnitudes. For those who adopted the Greek mathematical worldview, arithmetic was incomplete given the existence of incommensurable magnitudes. The acceptance of irrational "numbers" would be a much later development in the history of mathematics, and was never accomplished anywhere in the medieval period. This would await the advances made in the early Renaissance by such European figures as Christoff Rudolff, Giralmo Cardano, Johannes Scheubel, Rafael Bombelli, and others (see Volume 3).

Representative of the early medieval acceptance in the West of this Greco-Roman inheritance were views of St. Augustine, who held that mathematics was essential for the analysis of grammar, astronomy, and the other arts in general. Grammar, for example, depended on mathematics to measure lengths of sounds and syllables, and to define meters in poetry. In *On Christian Teaching*, Augustine valued mathematics for interpreting scriptural references and for the necessary part it played in anything requiring enumeration. Creation was inherently mathematical, and the devil was said to be incapable of making anything "according to number and measure." As Bruce Eastwood has observed, "Combining these justifications for the study of number and arithmetic with the very practical counting and arithmetical techniques of Roman primary schooling, early-medieval mathematics had no need or desire to go further beyond the knowledge available in Boethius' *On Arithmetic* (based on the Greek *Arithmetic* of Nicomachus, *c*. 100 CE) than the practicalities of finger-reckoning and an ability to manipulate the abacus" (Eastwood 2013: 318–19).

By the twelfth century, as scholars in Spain and Italy began translating Greek and Arabic works, an early largely forgotten figure today was John Philoponus. Educated in Alexandria, he was a Byzantine critic of Aristotle, noted philologist (also known as John the Grammarian), and a convert to Christianity. Philoponus wrote influential critiques of Aristotle, notably of the *Physics*, and not only opposed the Aristotelian theory of impetus, but advanced an early theory of inertia. He also argued strongly against arguments both Aristotle and Proclus had made for the eternity of the world (as had Simplicius). Often quoted by Galileo, his ideas would no doubt have been more influential, sooner, had Philoponus not been condemned by the early Church as a Christian heretic for views he held concerning the nature of the Trinity. Many of his works were translated into Arabic and were influential in Islamic circles, especially his studies of projectile motion and arguments against the possibility of an actual

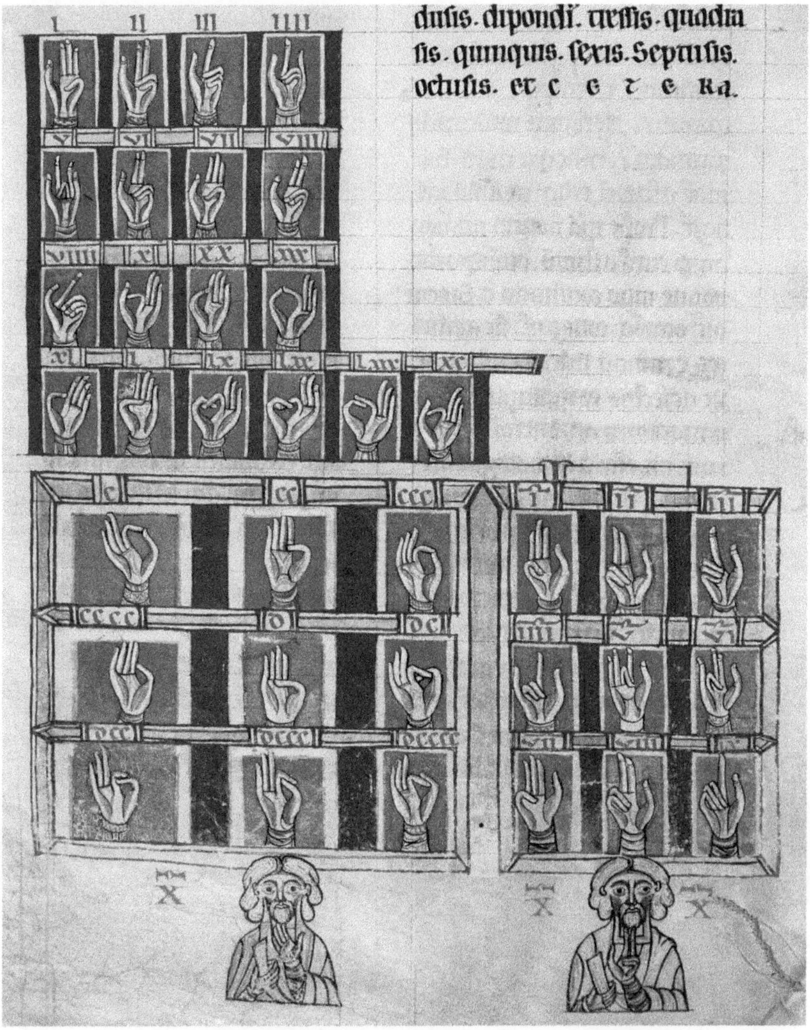

FIGURE 4.5. Finger counting from 1 to 20,000, from *De numeris* (On numbers). *Codex Alcobacense*, by Rabanus Maurus (780–856 CE), folio 251 verso. Germany, ninth century, now in the Biblioteca Nacional de Portugal, Lisbon. Courtesy of the National Library of Portugal. Photo by DeAgostini / Getty Images.

infinity and proof he offered for the temporal creation of the world. Although many accepted his arguments against impetus in favor of an inertial view of motion, al-Fārābī wrote a treatise against Philoponus, accepting instead the Aristotelian belief in the eternity of the world, as did Ibn Sīnā (Avicenna), Ibn Bājja (Avempace), and Ibn Rushd (Averroës). Nevertheless, Philoponus' ideas later proved especially influential in the medieval Latin West, and even later, for Galileo. As Stanley Jaki has observed:

> There is something tantalizingly revolutionary in the resolve by which Philoponus stated in his commentary on the *Physics* that all bodies would move in a vacuum with the same speed regardless of their weight (mass); that bodies with different weights falling from the same height hit the ground practically at the same time; that projectiles move across the air not because the air keeps closing in behind them, but because they were imparted a certain "quantity of motion." Today Philoponus' name is closely tied to a major recognition by twentieth-century history of science that Galileo had important and indispensable medieval forerunners, that the principles of momentum and inertia of moving bodies did not emerge during the seventeenth century as a *Deus ex machina*. (Jaki 1974: 186)

John of Sacrobosco (also known as John of Holywood), after studying at Oxford, continued his studies in Paris and eventually taught mathematics at the University of Paris. Through his teaching and writings, like *Algorismus* (also *De Arte numerandi*, first printed without date or place *c.* 1490), he introduced the new Arabic methods of arithmetic and algebra, including Hindu-Arabic numerals. In his major work, *Tractatus de Sphaera* (Treatise on the Sphere, 1220), Sacrobosco promoted the basic worldview of Ptolemaic astronomy, that the earth was spherical within a spherical universe. This was the first work on astronomy to be printed (in 1472). Sacrobosco was among the early proponents of calendar revision, having found the Julian calendar to be incorrect by about 10 days. His solution to updating the calendar was to delete one day every 288 years.

Nicole Oresme, after studying the arts and theology at Paris, became the dean of the Rouen Cathedral and later was appointed Bishop of Lisieux in 1377. Best known for his translations of the works of Aristotle, in his own work, *Livre du ciel et du monde* (Book of Heaven and Earth, 1377), he contemplated the possibility of a rotating earth, which he argued would simplify explanations of observed heavenly phenomena, like the diurnal rotation of the fixed stars. He even held an early notion of inertia, arguing that if the earth were moving, then so, too, were the other elements, for example water and air, thereby answering the objection that if the earth were moving, surely anyone would feel the wind such motion should cause. He also held that Scriptural passages referring to the apparent motion of the sun should not be taken literally. Although he appreciated the logical force of such arguments, he nevertheless in the end believed that the heavens actually did rotate around a fixed immobile earth at the center of the cosmos.

In terms of mathematics, Oresme's most significant innovation was his contribution to mathematical physics. The Aristotelian worldview had promoted a stark dichotomy between the presumably perfect heavens and imperfect earth—the terrestrial realm being one of constant change. On the other hand, the Neoplatonist view that mathematics was about an unchanging world of eternal truths precluded its application, or so the argument went, to anything that was forever changing. Projectiles, for example, always seemed to follow a different trajectory, just as the speed of a falling body was always changing, never moving at the same speed but always accelerating as it continued to fall.

Originally discovered by William Heytesbury and used by others associated with Merton College, Oxford, like Thomas Bradwardine and Richard Swineshead, what came to be known as the Mean-Speed Theorem (MST) (sometimes referred to as the "Merton theorem") was first proved graphically by Oresme. It was Oresme who discovered a means of approaching problems of constant change (in what was usually referred to as the intension and remission of forms, like changes in heat or the brightness of light) in a way that could be represented graphically and analyzed mathematically. This was possible so long as the rate of change in heat, for example, or whatever the form in question might be, was uniform. In the case of motion, for an accelerating body whose velocity is always changing, the MST states that if starting from rest and accelerating uniformly, the body will travel the same distance as one moving with a constant mean-speed whose velocity is one-half the final velocity of the uniformly accelerating body. This had the effect of reducing the case of a form that was constantly changing into one that was equivalent but involved no change, and thus was amenable to mathematical treatment.

Oresme did not apply the MST to the case of the uniform acceleration of falling bodies, a result first asserted but not proved by Domingo de Soto in his *VIII libros physicorum* (Eight Books on Physics, 1545), an influential commentary on Aristotle's *Physics*. Although Oresme and de Soto were thus forerunners of Galileo in this regard, Edward Grant summarizes the very significant difference between their achievements and Galileo's:

> Although Soto believed that uniformly accelerated motion is illustrated in naturally falling bodies, he did nothing more. By contrast, Galileo pursued the matter with great vigor and genius. He brought together all the significant concepts, definitions, theorems, and corollaries on motion and arranged them into a logical and ordered whole, which he then applied to the motion of real bodies. Uniform acceleration was no longer a mere hypothetical definition, but a true description of the way bodies fall in nature, as exemplified by Galileo's famous inclined-plane experiment. Galileo constructed a new science of mechanics and thereby laid the foundations of modern physics. (Grant 1996: 103–104)

Of all the medieval thinkers of the Latin West, Nicholas of Cusa (Nicolaus Cusanus) is regarded as among the most influential thinkers of the fifteenth century. Educated at Heidelberg and Padua, he was ordained a priest in the 1430s and was later appointed a Cardinal by Pope Nicholas V in 1448. In 1450 he was named Bishop of Brixen (Tirol), but later retired to Rome where he continued to serve the Church as an advisor to Pope Pius II. In his best-known work of 1440, *De docta ignorantia* (On Learned Ignorance), Cusanus deals with the created universe, which he took to be in a constant state of flux. This led Cusanus to some remarkable conclusions based not on empirical observation but on metaphysical arguments. The earth could not be fixed in one place because nothing was ever completely at rest. Because the universe is constantly in motion without any fixed center or finite boundaries, none of the spheres in either the Aristotelian or Ptolemaic world picture could be exactly spherical. Since the "outermost sphere" had no fixed boundary, the universe must be infinite, physically unbounded, potentially but not actually infinite. Cusanus thus prefigured the idea of an "infinite" universe, but one that was a limited image of God's actual infinity.

Cusanus was among those who adopted a strong form of Neoplatonism. In his view, numbers and mathematics were intimately connected with the creation itself—and no knowledge, he believed, could be more certain than mathematics, given that it was a construct of the human mind, itself created by God and a reflection of God's thought reflecting the underlying reality of the created world. For Cusanus, the certainty of mathematics was guaranteed by the fact that everything is created according to the divine word or *logos*, the immediate word of God. When Cusanus argued that the geometry of the sphere was such that its center was everywhere and its circumference was nowhere, this again reflected his view that the unbounded physical universe was an image, although necessarily limited, of God's actual infinity. In this sense Cusanus was reflecting the well-known words of the *Book of Wisdom* 11:20, that in creating the world, God had ordered "all things in measure and number and weight." As Cusanus maintained:

> In creating the world, God used arithmetic, geometry, and likewise astronomy. (We ourselves also use these arts when we investigate the comparative relationships of objects, of elements, and of motions.) For through arithmetic God united things. Through geometry He shaped them, in order that they would thereby attain firmness, stability, and mobility in accordance with their conditions. Through music He proportioned things in such a way that there is not more earth in earth than water in water, air in air, and fire in fire, so that no one element is altogether reducible to another. As a result, it happens that the world-machine cannot perish. And so, God, who created all things in number, weight, and measure [Wisdom 11:21] arranged the

FIGURE 4.6. God the geometer, from the *Codex Vindobonensis* (Osterreichische Nationalbibliothek), an ornately illustrated Bible (*c.* 1220–1230 CE), in this case showing the creation of the world, mathematically, according to carefully reasoned geometrical proportions. Courtesy of the Austrian National Library, Vienna / Wikimedia Commons.

elements in an admirable order. (Number pertains to arithmetic, weight to music, measure to geometry.) (Hopkins 2001: vol. 1, 99)

Just as Oresme prefigured Galileo, so, too, these views of Cusanus anticipated the resurgence of Pythagorean and Platonic ideas that were to play such significant roles in the Scientific Revolution, a revolution that would depend squarely upon mathematics.

CONTINUITY, ATOMISM, AND METAPHYSICS

Worldviews concerning the nature of matter and the material world necessarily involve mathematics with respect to debates over the nature of continuity, the discrete versus the continuous, and atomism. These in turn involve the problem of mathematical infinity, and whether matter, space, and time, for example, are infinitely divisible or ultimately comprised of some sort of very small, possibly infinitesimal but non-zero particles. Remarkably, such questions, while prominent elsewhere, were not widely if ever considered in medieval China.

From the earliest times, Chinese natural philosophers had reduced the basic elements of which all things were comprised to a Pythagorean sort of dualism between *yin* and *yang*, or an elemental theory of matter based upon five primordial elements or phases, namely metal, wood, fire, water, and earth, each understood as combining and separating in terms of recurring cycles. In general, Chinese thinkers never seriously developed an atomic theory of matter. Needham explains the reasons for this apparent lack of interest as follows:

> Atomism is one of the most familiar features of European and Indian theorizing, yet although at various times some Chinese thinkers watered its seeds, the idea never took root among them, presumably because it was out of harmony with those organic presuppositions on which Chinese thought was based . . . Chinese physical thinking was dominated throughout by the concept of waves rather than of atoms. (Needham 1962: 3)

A more subtle explanation is provided by Jean-Paul Reding, who sees atomistic theories as intrinsically linked to the problem of change. Beginning with Parmenides, Democritus, and Leucippus among the Greeks, Reding regards atomism in terms of the problem of *ex nihilo nihil* (nothing from nothing):

> [N]one of the Chinese words for change has concomitant representations similar to the Greek idea that a thing grows out of nothing or vanishes into nothing. Most of the Chinese terms, on the contrary, describe change as a rearrangement of pre-existing elements, and none of them suggests that a thing vanishes into nothing or comes out of nothing. (Reding 2004: 124–5)

Needham also related the lack of interest in atomism in China to the fact that in the Chinese worldview, there was neither a sense of "laws of nature" nor a belief in a "juridical" lawmaker. As sociologist and historian of science Toby Huff explains:

> In place of Western atomism governed by laws of nature, or the Islamic occasionalism governed by God's will, ancient Chinese thought was committed to an organic conception of the universe, suffused with primary forces such as the *yang* and *yin*, and the five elements ([*wu hsing*] wood, water, fire, earth and metal), constantly shifting in recurrent cycles. Within this cosmos there is no prime mover, no high god, no lawgiver. It is assumed that there is a pattern to existence and that there is a unique way (*tao*, or *li*, pattern) for all things, but these are embedded in recurrent cycles or phases. (Huff 2017: 205)

Atomism and the Infinite in Indian Thought

In India atomism reaches back to the earliest Vedic writings, reflected in the words of Uddālaka Āruni (early first millennium BCE): "particles too small to be seen mass together into the substances and objects of experience" (McEvilley 2012: 151). Within the next few centuries, elemental views of matter were proposed among Jains, Ājīvikas (an ascetic philosophical sect embracing an extreme form of determinism), and Lokāyatas or Cārvākas (radical materialists rejecting beliefs in an afterlife, reincarnation, or deities). Generally, these philosophies held that a few (four to seven) elemental substances are eternal, and out of the combinations of substance all things, including mind or soul, arise. The structure of matter among Jains and Buddhists was considered to be atomistic; it could not be infinitely divisible without making the concept of relative size meaningless, since all objects of any size would consist of equally infinite multitudes of parts.

Especially in Jainism, potential paradoxes about the size of the indivisible part or *paramāṇu*, occupying a single space-point, led to an enduring fascination with understanding the very small and very large. From at least the early first millennium CE, Jain doctrine classified some integers (if greater than unity, which was not considered a number) as enumerable (finite) or endless (infinite). But others partook of a different, intermediate category, called "innumerable" as distinct from "infinite"; in fact, they might be "basic-innumerable," "increased-innumerable," or "innumerably-innumerable," and each of those degrees could have a minimum, maximum, or intermediate magnitude. (An infinite quantity could likewise be basic, increased, or infinitely-infinite, but there was no maximum magnitude admitted for that last degree.) Computational and notational speculations of great delicacy and complexity were invoked to try to articulate the quantitative boundaries of these distinctions (Petrocchi 2017).

Other atomistic traditions, if less engrossed by the challenges of comprehending and expressing number, were equally engaged with the problems of interpreting matter as aggregates of imperceptible yet indivisible parts. The Nyāyá-Vaiśeṣika school of philosophy envisioned elementary atoms of numerous types, accumulating in regular combinations. By the seventh century the Buddhist philosopher Dharmakīrti, like Dignāga before him, advanced a theory regarding atoms as points of energy.

Paradoxes of the Infinite and Islamic Atomism

Islamic thinkers were greatly influenced by both Greek and Indian atomic theories, but they also had to confront certain controversial issues raised by religious orthodoxy. The most prominent group of Islamic atomists was the Ash'arite school associated with al-Ghazālī. For the Ash'arties, only atoms had an eternal, material existence. All else was considered "accidental," existing only from moment to moment. This was the basic occasionalist point of view to which al-Ghazālī adhered. Among the most prominent opponents of such views were the Aristotelians, above all Averroës, whose views were especially influential in later Jewish and Christian circles.

Among the atomists was Abū al-Hudhayl al-'Allāf, whose views were hotly debated by al-Naẓẓām, who invoked mathematical arguments against atomism, like the following. Suppose atoms to be dimensionless points, then given any atom on the diagonal of a square, a line perpendicular to the line at that point will join the atom to two others perpendicularly opposite each other on adjacent sides of the square. Every atom on the diagonal can be joined to a dimensionless atom on each side of the square, perpendicularly opposite the diagonal. This means there must be the same number of atoms on the diagonal as on any two adjacent sides of the square. Assume the atoms to be moving uniformly, then it must follow that the time it would take for one point to traverse the diagonal must be the same as that for another point to traverse two adjacent sides of the square. The obvious contradiction here led to the rejection of Avicenna's material atomism. This argument also appeared in Muḥammad al-Ghazālī's *Tahāfut al-Falāsifa* (Incoherence of the Philosophers), and when translated into Latin was thereby particularly influential in later medieval European thinking about such matters.

Moses ben Maimon (Maimonides) (1138–1204), the Sephardic Jewish astronomer, philosopher, and physician, also used mathematical arguments to counter the atomists, in particular the Mutakallimūn (Kalamists):

> Nor must you suppose that the aforegoing theory concerning motion is less irrational than the proposition resulting from this theory, that the diagonal of square is equal to one of its sides, and some of the Mutakallemim go so far as to declare that the square is not a thing of real existence. In short, the adoption of the first proposition would be tantamount to the rejection of all that has been proved in Geometry. The propositions in Geometry would, in

> this respect, be divided into two classes: some would be absolutely rejected, e.g., those which relate to properties of the incommensurability and the commensurability of lines and planes, to rational and irrational lines, and all their propositions in the tenth book of Euclid, and in similar works. Other propositions would appear to be only partially correct; e.g. the solution of the problem of dividing a line into two equal parts, if the line consists of an odd number of atoms; according to the theory of the Mutakallemum such a line cannot be bisected. (Maimonides, *Guide for the Perplexed*, quoted from Bell 2006: 53)

Medieval Atomism in the Latin West

Given that in the medieval Latin West the ideas of Aristotle dominated thinking in natural philosophy, it is no wonder that little credence was given to an atomic view of the material world. Occasionally some works did offer an atomistic approach, as in writings attributed to Pseudo-Geber, the unknown author of several alchemical treatises of the late thirteenth and early fourteenth centuries. Said to have been written by Abū Mūsā Jābir ibn Hayyān, an alchemist of the Golden Age of Islamic science, the major work of this corpus, *Summa perfectionis magisterii* (The Height of the Perfection of Mastery), was influential primarily for alchemists and metallurgists, and advanced a theory that all matter was comprised of minute particles of mercury and sulphur.

Somewhat later, Nicholas of Autrecourt opposed the views of both Aristotle and Averroës and instead favored an atomistic theory that accounted for changes in the natural world in terms of the motion of infinitely small particles of matter. These views complemented his arguments in favor of the eternity of the world and the infinite divisibility of space and time, both of which he viewed as comprised of indivisible points of space and instants of time. The similarities between his ideas and those of al-Ghazālī suggest that he may have been familiar with Averroës' arguments against al-Ghazālī (Marmura 1968: 286–9). However, Autrecourt's ideas were to have little impact due to the condemnation of his teachings at the University of Paris by the papal court in Avignon in 1346. Only when Aristotelian thinking began to be challenged in the wake of translations of Lucretius' *De rerum natura* (On the Nature of Things) in the early fifteenth century did consideration of atomism stir any interest among medieval natural philosophers, paving the way for the later atomistic theories of Pierre Gassendi and Robert Boyle, among others.

One significant factor that contributed greatly to the Scientific Revolution, in fact, was the influence of the corpuscular philosophy. The significance of the rejection this entailed of the prevailing medieval worldview has been summarized succinctly by David Lindberg:

> In exchange for the purposeful, organized, and (in many ways) organic

> world of Aristotelian natural philosophy, the new metaphysics offered a mechanical world of lifeless matter, incessant local motion, and random collision. . . . For the explanatory capabilities of form and matter, it offered the size, shape, and motion of invisible corpuscles. . . . And for Aristotelian teleology, which discovered purpose within nature, it substituted the purposes of a creator God, imposed on nature from the outside. . . . And finally, the shift of emphasis from Aristotelian qualities to the geometrical properties of corpuscles (shape, size, and motion) surely encouraged the application of mathematics to nature. (Lindberg 1992: 361–2)

As Lindberg concludes: "This produced a radical conceptual shift, which destroyed the foundations of natural philosophy as practiced for nearly two thousand years" (ibid.: 361). Instead of a terrestrial worldview that found no essential role for mathematics in a realm of constant change, one of the distinguishing features of mathematics in the Renaissance was the new ways it found to make mathematics applicable to analyzing and predicting a wide range of natural phenomena. How this might be done eluded mathematicians in all parts of the world during the medieval period. While mathematicians in China, India, and the Islamicate world had made numerous important discoveries and pioneered in algebra and trigonometry in particular, the power and scope of mathematics was about to expand dramatically in the course of the Scientific Revolution in Western Europe, and above all through the advances made possible by analytical geometry and the infinitesimal calculus in particular.

CHAPTER FIVE

Describing and Understanding the World

Medieval Mathematics of the Heavens and Earth

JIAN-PING JEFF CHEN

Records of mathematics, printed or otherwise, are fundamental primary sources for the study of the histories of mathematics. As evidenced in the previous chapters of this volume, traditional mathematics in the medieval period, in different parts of the world, followed different cultural and textual traditions, manifested in different styles that sometimes varied dramatically in content while sharing certain common elements. Likewise, the utility of mathematics and the roles it played in different medieval cultures also reflects important similarities across differing geographical locations. In this chapter, the cultural history of traditional mathematics is explored primarily in four regions during the medieval period: China, India, Islamicate countries, and the Latin West. Although mathematical activities and their cultural influences may be viewed from the perspective of how mathematics was used in service to ruling classes, in the cultivation and pursuit of the intellect, and in a wide variety of popular applications and practices common to everyday life, we will focus in what follows primarily on how mathematics contributed to describing and understanding the natural world through astronomical modeling and astrological prediction (in particular through trigonometric techniques), cartography and mapmaking, land surveying, and where applicable, optics and actuarial problems, including games of chance.

UNDERSTANDING THE WORLD MATHEMATICALLY IN MEDIEVAL CHINA

In China, the concept of *suan* (literally, "calculation") arguably best described the activities in the culture that would be considered "mathematics" from a modern point of view. Unsurprisingly, procedures of calculation held a prominent place in the records of mathematics, wherein collections of question-answer-procedures formed the major component of the content of Chinese mathematics. Many treatises also contained commentaries either written by the authors themselves or amended by scholars in later generations. As there was no clear professional identification that distinguished between mathematics and astronomy, these disciplines shared the same practitioners in medieval China. Thus the model of recording computational knowledge in the form of algorithmic procedures that typified medieval Chinese mathematics can also be observed in astronomy.

The first and foremost example of mathematics in the service of statecraft concerns the computations at the heart of mathematical astronomy. Traditionally, mathematical astronomy was an integral part of calendrical science, which included calculating and predicting the dates and times of solar and lunar eclipses, and the positions of the planets, going far beyond issuing an annual almanac for the empire. Above all, a Chinese calendar demonstrates the heavenly mandate, and thus was crucial for validating the legitimacy of a dynasty.

One Han dynasty astronomer, Liu Xin, described the diverse utility of mathematics not only for astronomy, but for understanding the world generally, as follows. "In deriving astronomical systems, producing pitch-pipes and making vessels, encompassing the circle and setting right the square, weighing the heavy and balancing what is equal, leveling the builder's line and making fair the capacities, in seeking the profound and drawing out the obscure, hooking up from the deep and attaining to the distance, there is nothing in which numbers are without applications" (Cullen 2009: 603–4).

The Chinese calendar has always been a luni-solar one. A major component in calendrical computations is reconciling the incommensurate periodicities of the motions of the sun and moon. Moreover, the lengths of these periods ultimately define the length of a year and a month. Until the Han dynasty, the Chinese calendar was based on a "quarter remainder system" in which a solar year was taken to be 365 ¼ days. In 104 BCE the *Taichu li* (Grand Inception Calendar) was introduced with a lunar cycle of 29 43/81 days and a solar cycle of 365 385/1539 days. Later calendars relied upon improved numerical values for the lengths of a month or year. For practical purposes, a lunar month in the Chinese calendar has either 29 or 30 days in order to accommodate the non-integer length for a single lunation. A year in the calendar is comprised of 12 lunar months which fall short of the 365 ¼ days in a tropical year. To reconcile the difference, a general rule was adopted, that every few years an intercalary (lunar) month would be inserted in the calendar. In the course of time, the lengths of a month or year in later sources were often more "accurate." For example, in the *Dynastic History of Jin* (266–420 CE), the number of days in a year was given as 365 145/589 and in a lunar month as 29 773/1457.

In the official dynastic history of any period, there is always a section devoted to mathematical aspects of musical instruments and the making of calendars, e.g., a "Treatise on Pitch Pipes and Astronomical Canons," in which the portion concerning the calendar is mostly comprised of procedures (this work has also been referred to in English as "Monograph of Musical Scales and Astronomical Systems" (Cullen 2016: 365). A typical treatise of this kind often contains an historical account of the evolution of prior calendars as well as procedural texts for a specific *li* or calendar. In particular, many astronomical constants are listed and algorithms given with step-by-step procedures to calculate the "precise" moments of important occasions such as an astronomical new moon or the exact time of a winter solstice. In a system wherein procedural texts were recorded as part of such official histories, even government functionaries who had little understanding of the theory behind the algorithms could perform the computations to make astronomical predictions, and they had no need to be engaged in grand discussions of cosmology or cosmography. For details, see Martzloff (2016).

Basically, there were three cosmographical models in China. The *gaitian* model, seen as early as the *Mathematical Classic of the Zhou Gnomon*, composed *c.* first century BCE, wherein the sky was treated as a hemispherical canopy overarching the flat earth. The *huntian* model, developed even earlier during the Warring States Period (475–221 BCE), considered the sky as a circular sphere with a square earth suspended within, sometimes compared to the egg yolk inside an egg, but flat. Considerably later, a third theory, the *xuanye* model, appeared in the second century CE and is mentioned in later dynastic histories, including those of the Jin and Sui dynasties. This theory did not specify the shape of the universe but considered it to be infinite with celestial bodies hanging within. However, no known calendar system was based on the *xuanye* model in making astronomical or calendrical computations. For later calendrical developments towards the end of the medieval period in China, see the closing section of this chapter.

MATHEMATICS AND MAPMAKING

Mathematics and Mapmaking in China

Mathematics was equally crucial to the development of accurate terrestrial maps in medieval China, and had long been of use for administrative, political, and military purposes. In one famous incident during the Warring States Period, a map was even used by an assassin who gained an audience with the King, who later became the first emperor of the Qin dynasty by presenting a map of the Kingdom of Yan in which a fateful dagger was secretly concealed.

In 1986, a number of maps drawn in ink, on four wooden boards, were found in the tomb of a Qin dynasty army officer (*c.* 239 BCE) at Fangmatan in Gansu province. The Fangmatan maps depict regions of strategic importance, as do maps from somewhat later found in 1973 at Mawangdui near today's city of Changsha, the capital of Hunan province. These maps, three on silk, were buried in 168 BCE in the tomb of a prominent official of the Changsha state. The maps are detailed and in color, with rivers and county seats identified by name. One is a topographical map, one a military map, and the third is the plan of a prefecture.

Areas of military significance are clearly noted, including army installations, headquarters, roads, and settlements in red, streams and rivers in blue-green, other features and lettering in black ink. Distances between some settlements are noted and the number of households is also given. These may reflect scale mapping and some have pointed to a note in the early *Mathematical Classic of the Zhou Gnomon* wherein reference is made to adopting a scale of 1 *fen* for every 2000 *li* (a scale of roughly 23.1 to 321,600 meters, or roughly 1:36,000). Several centuries later, one of the earliest texts on mapmaking written in the third century CE by Pei Xiu also stressed the importance of careful measurement and scale for accuracy in representing the details of any geographic landscape.

Five hundred years later, the Tang dynasty general Jia Dan was renowned for his mapmaking skills. A prominent scholar-official credited with producing maps of high quality, historical records indicate that Jia Dan made a map of China along with neighboring countries entitled "Map of Chinese and non-Chinese Territories in the World," now lost. According to Joseph Needham, "upon its completion in 801, the map was 9.1 meters in length and 10 meters in height, mapped out on a grid scale of one inch equalling one hundred *li*" (Needham 1959: 543)

Marginal notes on another map of 1136, the *Huayi tu* (Map of Chinese and foreign lands), indicate that the names of foreign countries inscribed on its edges were taken from Jia's map. This map identifies various rivers and tributaries, four lakes, ten mountain ranges, and includes about five hundred place names, with notes providing additional information on lands beyond China.

The *Huayi tu* is engraved on a stele now preserved in the Forest of Stone Steles Museum in Xi'an, China. On the opposite side of the *Huayi tu* is another map, the *Yüji tu* (Map of the tracks of Yü), also engraved in 1136. It includes a grid of 5,110 squares, each representing an area of approximately 100 *li* (a scale of about 1:5,000,000). Notable features include the Shangdong peninsula, Hangzhou Bay, and Hainan Island, along with the names of more than 1,400 administrative units. This is the earliest map from China drawn with a cartographic grid, indicating the level of instrumentation and surveying techniques in China for making both direct and indirect measurements needed

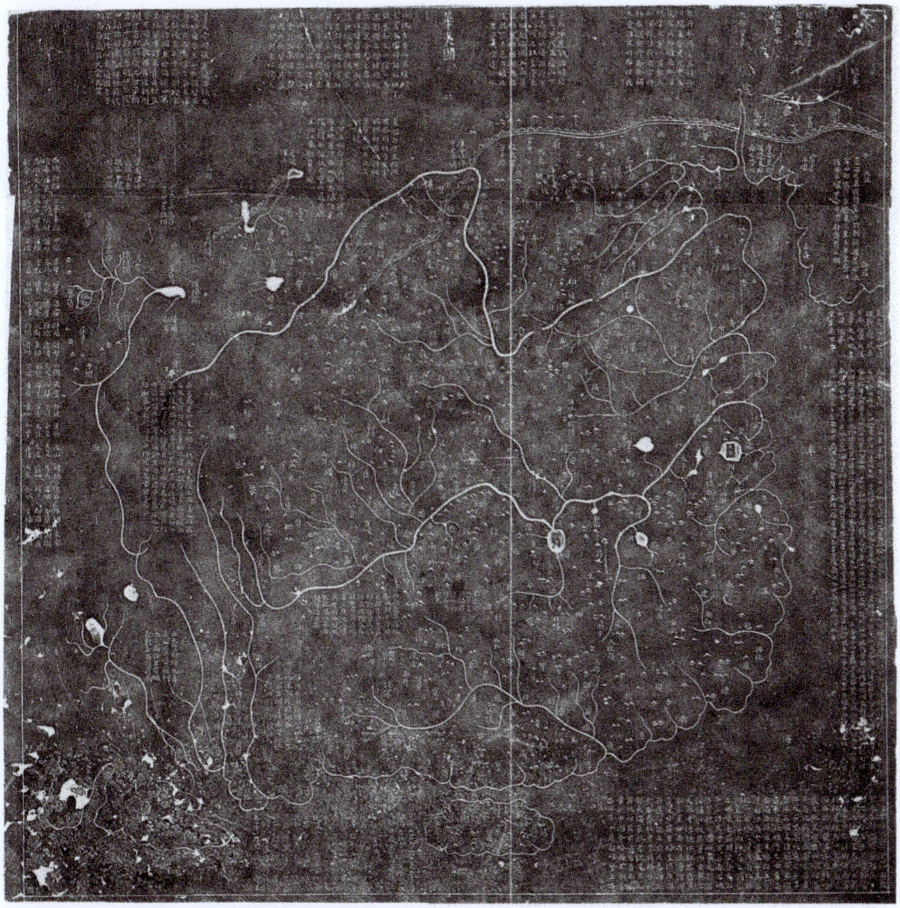

FIGURE 5.1. Stone rubbing of the *Huayi tu* (1136/1137 CE) map of China, engraved on stone, now in the Forest of Stone Steles Museum in Xi'an, China. This map represents Northern Song dynasty China, and includes the Great Wall in the north to as far south as Hainan Island; it identifies mountains, rivers, lakes, and more than 400 administrative place names. Photo by Universal History Archive / Universal Images Group via Getty Images.

to create maps with the accuracy reflected in the *Yüji tu*. These maps were believed to have been used to educate court officials about important locations and major topological features of the empire.

In the early Ming (1368–1644), the most influential map was doubtless "Zheng He's Navigational Map." This is believed to have been based on maps recorded during the first six voyages of the famous Chinese admiral Zheng He, who in all undertook seven voyages between 1405 and 1433, venturing as far west as the coast of Africa and the Arabian Peninsula. Evidence shows that the Zheng He map was most likely drawn after 1422 and before 1431 when he began his seventh and last journey. Each country visited during the first six journeys was recorded along with its distance from China. The earliest extant version of this map is preserved in the *Records of Armaments and Military Provisions* (1628), a military treatise composed by Mao Yuanyi.

Later maps in China also continued to make use of the grid system, as in those created by the Ming dynasty cartographer Luo Hongxian. Luo's *Guangyu tu* (Enlarged Territorial Atlas), printed in Nanjing, *c.* 1558, was based on an earlier map, now lost, the *Yudi tu* of the Yuan dynasty cartographer Zhu Siben. Maps in the *Guangyu tu* collection are all drawn with grids to scale. Some maps represented all of China, from the eastern coastal regions to as far west as Xinjiang province and Guangdong province in the south; others mapped regions beyond China, including Japan, Korea, and Vietnam, in addition to other parts of Southeast Asia.

In all, Luo's Atlas includes forty main maps, including sixteen provincial maps and nearly seventy minor maps. Among information recorded on the maps is data about defense structures and military posts along the borders and coastal regions, with data giving the numbers of soldiers and horses posted at border towns, as well as information regarding ethnic groups in the border regions as well. Some maps delineated the Great Wall of China, indicating its extent over more than two hundred miles, including more than a hundred fortresses and a thousand beacon towers used to signal danger from invading Mongolian nomads. Luo's atlas was later used by the Italian Jesuit Martino Martini in producing his *Novus Atlas Sinensis* (1655), which remained the authoritative source for Western maps of China until Jean Baptiste d'Anville's *Atlas de la Chine* published in 1737.

Mathematics and Mapmaking in India

Despite knowledge of the topography of India extending back to ancient times, there is an unfortunately conspicuous lack of actual maps originating in India during the medieval period. This is no doubt due to the combination of climate and the perishable materials on which most Indian documents were kept—written primarily on palm leaves, paper, cloth, and other organic materials—these were easily subject to deterioration and subsequently lost to the historical record.

Before the appearance of Islam, India was what has been termed a "cosmically attuned society," one for which terrestrial maps held little interest. This may be reflected not only in the lack of terrestrial maps, but in a lack of narrative history prior to late medieval times. Thus prior to the thirteenth century, what is known of India is largely the result of the reports of foreign travelers. Likewise, the predominantly oral nature of Indian learning was not conducive to cartography. A basically oral tradition depending on memory was not one that lent itself to the creation of graphic images. The specific information required for useable maps, including detailed records of longitude, latitude, the position of coastlines and major landmarks and any prominent locations requiring identification, would have been difficult to commit to verbal memory alone. As one of the world's leading experts on the subject of Indian cartography puts it, much of the problem also lay with the religious insignificance of this world as opposed to the next:

> In short, then, to those of a religious bent – which for many centuries probably included most learned persons – so mundane a task as preparing a seemingly accurate map of the finite terrestrial earth or a small segment of it could not have appeared particularly important. By contrast, making

FIGURE 5.2. Al-Idrīsī's world map of 1154 CE, from a copy made in 1456 by Alī ibn Hasan al-Hūfī al-Qāsimī at Cairo (Bodleian Library, MS. Pococke 375 fol. 3v-4.) Courtesy of the Bodleian Library, University of Oxford, Oxford, UK / Wikimedia Commons.

> manifest the structure of the cosmos, through which the soul makes its long cosmic journey, must have been regarded as an act of considerable importance. To the devout, the diverse cosmographies put forward provided, as it were, a choice of road maps for the soul. In comparison with the transcendent importance of this need, the necessity for route maps of a more mundane nature was probably not very keenly felt. (Schwartzberg 1992: 329)

Mathematics and Mapmaking in Islam

By the ninth century, under the patronage of the ʿAbbāsid Caliph al-Ma'mun, geographers had corrected earlier maps that had depicted the Indian ocean as a land-locked sea, as Ptolemy had imagined it to be. In the eleventh century the Persian geographer Abū Rayḥān al-Bīrūnī spent considerable time in Northern India. Although there is no evidence that he ever compiled a map of the country for his book, *Tarikh al-Hind* (Description of India, 1032), had he done so it has been suggested it would have surpassed any European map prior to the mid-sixteenth century. In 1025 he did write about the preparation of a world map for which he pinpointed the positions of many locations in India and throughout the Islamicate world by longitude and latitude. He also used travelers' accounts of the distances or travel times between cities that he listed in his book on astronomy, the *Kitāb al-qānūn al-Masʿūdi* (The Masʿūdi canon) (Schwartzberg 1992: 323–24).

MATHEMATICS AND TIMEKEEPING

Timekeeping in China

As dates and times were of crucial importance for setting court rituals, accurate timekeeping was indispensable for dynastic operations. In addition to using ordinal numbers to count the days in a month, another system was also in place, a sexagenary system using ten "heavenly stems" and twelve "earthly branches" (see Chapter 1 for further details of this system) whereby each day of the month was uniquely designated over a 60-day cycle, after which the names repeated again. From the end of the Western Han Dynasty (202 BCE–9 CE), this system was also used to designate years. However, this sexagesimal means of designating days and years was used primarily for purposes of divination and prognostication.

Mathematics was likewise critical in providing accurate means of timekeeping to measure the hours of the day and night. The same bureau responsible for astronomical observations and calendar making also managed timekeeping in general. The most elaborate of such devices was the creation of the Song dynasty polymath and statesman Su Song (1020–1101). In a treatise composed in 1092, "*Essentials of a New Method for Mechanizing the Rotation of an Armillary*

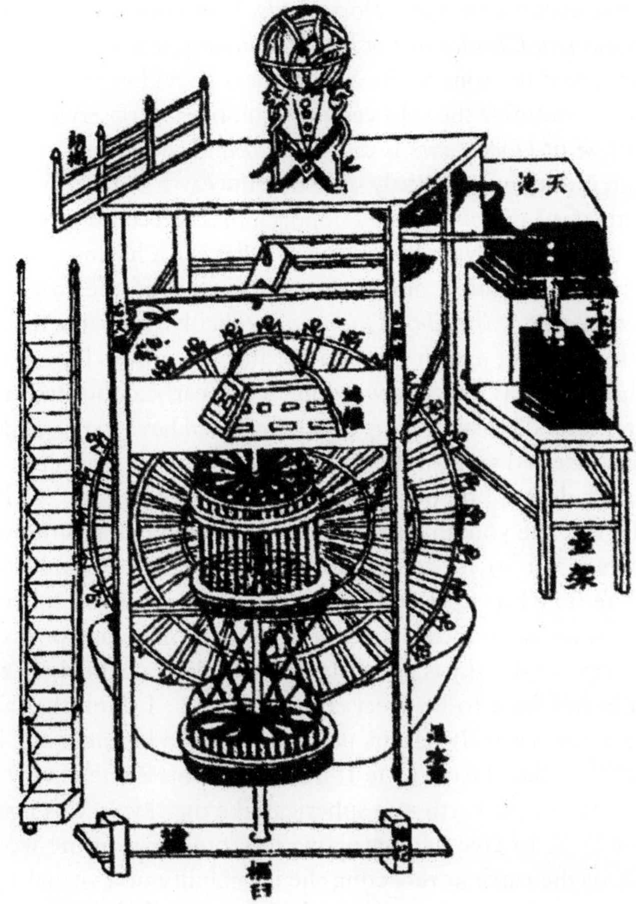

FIGURE 5.3. Schematic diagram of the inner workings of Su Song's clocktower built in the Northern Song dynasty capital of Kaifeng, from Su Song's book, *New Design for an Armillary Sphere and Celestial Globe* (1092). Photo by The Picture Art Collection / Alamy Stock Photo.

Sphere and a Celestial Globe," Su Song described in detail his water-driven mechanical clock tower, a device that employed an escapement mechanism for the accurate determination of the time (Needham 1965: 449–63). Other kinds of timekeeping mechanisms are also recorded, as for example in the Tang dynasty military treatise written by Li Quan, *Classic of the Bright White Yin* [*Star*] (sometimes translated as *The Scriptures of Venus and Yin*) and the Northern Song military encyclopedia, *Complete Essentials of the Military Classics*. However, as the methods they employed were basically different ways of relating counting to distances in military applications, we omit further discussion here.

The shape of the earth was never a controversial issue in China. The dominating view had always been that the earth was flat. In one of the most revered writings, from the Shang dynasty, *Honored Documents*, later known as one of the ancient classics as *Document Classics* or *Book of Documents*, it is said that the legendary sage, King Yao, sent the sons of the astronomers Xi and He to the four quarters of the world to determine the solstices and equinoxes by observing the stars. This part of the *Book of Documents* is dated to around the fourth or fifth centuries BCE. Although it does not explicitly state that the earth or the world was flat, the reference to the four directions in an important record could be easily construed by later scholars as employing the notion of a flat earth having come down from antiquity. The mathematical and astronomical text of the first century BCE, *Mathematical Classic of the Zhou Gnomon*, explicitly states that the sky is round and the earth is square, and in this context this has always been understood to mean that the earth was flat. The wording in this early authoritative text set the framework for later discussions; any dissenter would have to reconcile a different view with this received view of the shapes of the heavens and earth.

The *Huntian* description of the universe as an egg and the earth suspended in the middle like the yolk in an egg refers to the relative position of the earth in the universe, not to its actual shape. When the medieval Song dynasty polymath Li Ye (Li Zhi) expressed his view that the earth was spherical like the universe, his declaration in effect reaffirmed the dominance of the traditional view: "The shape of the sky is precisely a sphere, like a pellet ball; the shape of the earth does not have to be precisely a square . . . I think that although the shape of the earth is mostly flat, its periphery should be spherical, like the sky, only smaller" (translated from Li Ye 1985: 1, 1a). This is one of very few records that explicitly states the earth was spherical like the sky. In this instance it was necessary for Li Ye to give an alternative interpretation of the word "square" for the shape of the earth as reflecting the immobility of a virtual square which he believed the earth also shared in order to reconcile his view with the text from a classic. About this, Li Ye wrote as follows: "Moreover, the earth being immobile and not moving, manifests the virtue of the square, and [therefore] can also be described as square" (Li Ye 1985: 1, 1a). As Needham has noted, these individual claims for a round earth had little influence on Chinese cartography, which "revolves around the basic plan of a quantitative rectangular grid." The debate over a flat earth versus a spherical one in Chinese intellectual circles has to wait until after the Jesuits introduced the theory of a spherical earth to Chinese scholars in the seventeenth century, which falls outside the scope of this chapter.

Timekeeping in the Islamic Mideast

Second-century Egyptian engineers added a clever kind of self-regulation to water clocks. Their version was picked up by Arab artisans; and the Moors of medieval Spain finally brought clockmaking to a high state of development.

Among the earliest examples of the technological expertise Islamic craftsmen had achieved by the ninth century CE is the presentation of an elaborate water clock to the Holy Roman Emperor Charlemagne in 807 by the 'Abbāsid Caliph Hārūn al-Rashīd in Baghdad. According to the *Annales Regni Francorum* (*Annales Laurissenses Maiores*) in the year 807, the clock was "arte mechanica mirifice conpositum," a work of mechanical artistry "admirably arranged," such that the device was moved by a water clock with brass balls arranged to fall at the hour onto a cymbal, when a horseman would emerge from a window, twelve in all, one closing at the opening of another to signal the progression of hours. For a detailed discussion of the clock and its workings, see Al-Hassani (2016).

Much more advanced was the device designed by the engineer Muḥammad al-Sā'ātī's, who constructed a water clock for the Umayyad Mosque in Damascus, *c.* 1154. This remarkable instrument marked the time, by day, when a small crescent moved at a constant speed before doors which would open to announce the hours in succession, as two falcons would lean over to dispense copper balls into a vase, sounding the passing of each hour, and then returning to their upright positions. At night, a lamp inside indicated the hour with the help of a rotating disc. The clock was rebuilt by al-Sā'ātī's son, Riḍhwān ibn Rustam al-Sā'ātī, who wrote a treatise on clockmaking in 1203, the *Kitāb fi 'ilm al-sā'āt wa'l-'amal bihā* (The Science of Clocks and Working with Them), in which he described with illustrations the workings of the Damascus water clock.

Of the many timekeeping devices included in the *Kitāb fī ma'rifat al-ḥiyal al-handasiyya* (The Book of Knowledge of Ingenious Mechanical Devices) by Ibn al-Razāz al-Jazarī, "The Elephant Clock" is perhaps the best known. Al-Jazarī wrote his treatise early in the thirteenth century, and in addition to the elephant clock, this work includes detailed descriptions, fully illustrated, of a variety of remarkable mechanical devices, among them fountains, drinking vessels, and water clocks, along with devices for raising water. The elephant clock rang every half hour, when the bird on the dome would whistle while the figure below dropped a ball into the mouth of the dragon and the driver atop the elephant would drop his halberd to spur it on.

The most sophisticated of al-Jazarī's instruments was the "castle clock," which due to its intricate mechanism some have described as a programmable analog computer. This was a complex instrument nearly 33 meters (108 feet) in height, which in addition to telling the time also exhibited the sun and moon moving through the zodiac, with a crescent moon that moved above a series of twelve doors, each opening in turn to reveal a figure announcing the hour. It was reprogrammable day-by-day to account for the differences between the lengths of night and day as the year progressed, from the solstices through the equinoxes. Among the automata were five musicians designed to play music created by a camshaft moved by a water-wheel. Two falcons, overlooking the musicians, were programmed to drop brass balls into vases below them to sound the hours.

FIGURE 5.4. "The Elephant Clock," as depicted in the *Kitāb fī ma'rifat al-ḥiyal al-handasiyya* (The Book of Knowledge of Ingenious Mechanical Devices) by Ibn al-Razāz al-Jazarī (dated 715 AH/1315 CE). Photo by Sepia Times/Universal Images Group / Getty Images.

Later astronomers made further improvements, including an astrolabic clock constructed by Ibn al-Shāṭir for the Umayyad mosque in Damascus, where he was the leading *muwaqqit* of those responsible for determining the times of prayer each day. This required detailed mathematical calculations, which he helped to make more precise with the aid of a large horizontal sundial he placed atop the northern minaret of the mosque. This was made of marble measuring some 2 meters in length, with curves incised to read the hours for prayer at their appointed times between sunrise and sunset, all with a gnomon aligned with the celestial pole, an advance usually associated with later European astronomy (King 2007: 569–70). Constructed in 1371/72 atop the main minaret of the Damascus Mosque, the various lines on this instrument allowed the *muwaqqit* to tell the time not only in seasonal as well as equinoctial hours, but also relative to each of the five daily times for prayer. See as well (King 1983: 547) for a photograph and drawing of the device in question.

Further advances in the measurement of time were made by the Ottoman engineer Taqī al-Dīn, reflected in a book he wrote on the subject of clockmaking in 1563, the *al-Kawākib al-durriyya fī waḍ 'al-bankāmāt al-dawriyya*. He was the founder and director of the Constantinople observatory, begun under orders of the Sultan Murad III in 1569, for which Taqī al-Dīn devised clocks moved by weights using verge-and-foliot escapements producing as accurate readings as possible. Among them was a clock he was commissioned to design by the Grand Vizier that would tell the times when devout Muslims would be called to prayer.

Timekeeping in the West

Among European timepieces, the first mechanical escapement, the "verge escapement" (from the Latin for "rod," *virga*), did not make its appearance until sometime in the thirteenth century. Among the first to make use of the device were Richard of Wallingford (see Figure 2.1 in this volume) and Giovanni de' Dondi, both early in the fourteenth century. Richard undertook a thirty-year project to construct an astronomical clockwork for St. Alban's Abbey, just north of London, of which he served as Abbot. A University of Oxford graduate, he brought all of his knowledge of mathematics and astronomy to bear on the clockwork mechanism he so carefully designed, which included a twenty-four-hour strike mechanism, mechanisms to show lunar and solar eclipses, and detailed planetary gearing. It is the oldest purely mechanical clock for which accurate details are known, thanks to a manuscript he left behind now in the Bodleian Library at Oxford. Based upon Richard's notes, the Oxford historian of science John North was able to reconstruct the various parts on which Richard's mechanism depended, included a double-wheel verge and folio escapement, an astrolabe style dial indicating the positions of the sun, moon and stars, and a means of exhibiting not only phases of the moon but the corresponding ebb and flow of the tides.

Meanwhile, only a few decades later, Giovanni de' Dondi of Padua, Italy, completed construction of his famous Astrarium in 1364. Like Richard of Wallingford's astronomical clock, de' Dondi's instrument relied on an escapement as a crucial device, using a toothed wheel to control the uniform

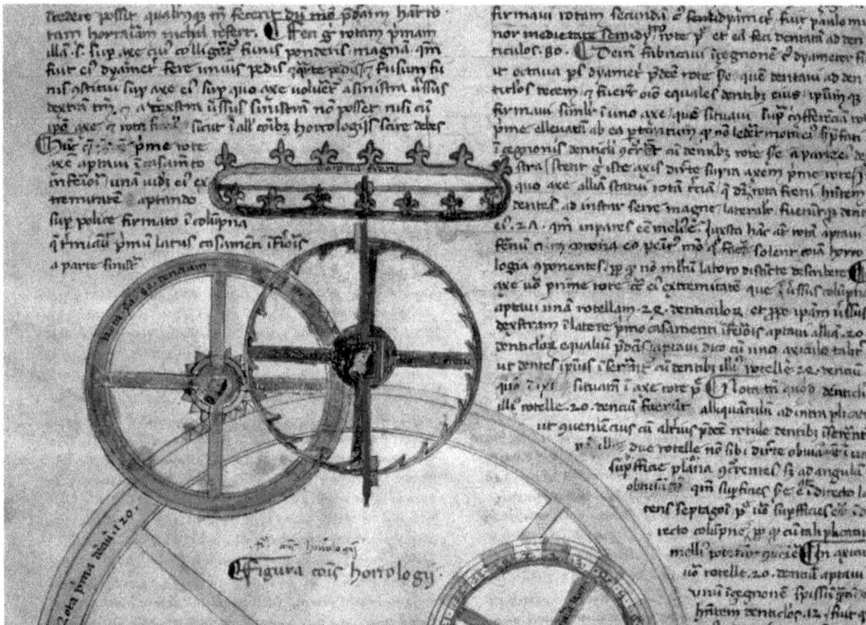

FIGURE 5.5. A page from Giovanni de' Dondi's *Tractatus astrarii* (1364), Biblioteca Capitolare, Padova, Italy. Courtesy of the Capitolare Library, Padova; photo by DeAgostini / Getty Images.

advance of the timepiece, all regulated by a verge or rod devised in such a way that as the verge released a tooth on the wheel, the wheel would then advance by a fixed and uniform amount until the verge would stop the movement, locking the escapement into place until the verge again released it.

De' Dondi's clock, based on the Ptolemaic geocentric model of the heavens, was comprised of seven faces, with 107 moving parts. In addition to five faces for each of the then-known planets (Mercury, Venus, Mars, Jupiter, and Saturn), there was also a face for the sun and another for the moon. It was designed not only to show the hours of the day, the days of the week, and the months as they progressed, but the solstices, equinoxes, and major religious holidays as well.

While de' Dondi's mechanism was only about a meter in height and was displayed for a time by Gian Galeazzo Visconti, the Duke of Milan, in the library of his castello in Pavia, its later fate is unknown. Larger astronomical clockworks were devised for public buildings or cathedrals and were on public display sometimes within, sometimes on exterior walls in major cities through Europe. The oldest such astronomical clock still in working order was built by Nikolaus Lillienfeld for the Straslund Cathedral in 1394 in what is today Straslund, Germany. Several decades later, the Prague Orloj (Prague Astronomical Clock) was built in 1410 for the Old Town Hall, and is also still keeping time in Prague, capital of the Czech Republic. Others that were rebuilt in post-medieval times, in chronological order of construction, include astronomical clocks in Strasbourg, France (1352–4); Rostock, Germany (1472); Auxerre, France (1483); Venice, Italy (1496); Tübingen, Germany (1510); and Münster, Germany (1540).

MATHEMATICS IN MEDIEVAL INDIA

Traditionally, the earliest recorded mathematics in India is found mostly in highly compressed Sanskrit of religious scriptures, which include astronomy, astrology, calendrics, and rituals. Sanskrit texts are valued mostly as a record of spoken language rather than a written one. The ancient mathematical texts presented as hymns in various linguistic styles include numbers, numerological ideas, the geometric shapes and sizes of fire altars, along with rules for laying out and building them. Here the mathematics was geometrical in nature and associated with sacrificial rituals. Except for the sexagesimal-base used for certain time-units and astronomical measures, the basic means of numbering in Indian mathematics used a decimal place-value system.

During the medieval period, mathematics can be found primarily in two sources: in parts of treatises on mathematical astronomy (astronomical *siddhāntas*) and in stand-alone mathematical works. In astronomical *siddhāntas*, the main purpose was to calculate the positions of the celestial bodies, and these were works composed in verse and typically included chapters on calculations (*gaṇita*). In the earliest completely preserved *siddhānta*, the *Āryabhaṭīya* of

Āryabhaṭa, the second chapter is devoted to calculation. The contents include arithmetical and geometric procedures, finding square or cubic roots, the areas of various geometric figures, calculations of series, and trigonometrical calculations of sine lengths, shadows, chords, and circles. It is worth noting that the *Āryabhaṭīya* was written as a presentation of methods of general calculations, and not to demonstrate astronomical applications. Later, commentaries were added to assist readers in understanding the proper interpretation of often very cryptic rules. Both the solutions and the commentaries assume a certain level of familiarity with basic arithmetic.

Unlike the problems in Chinese mathematical treatises, very few examples deal with concrete situations from daily life. Exceptions include the calculation of interest on a given amount of capital and problems involving horse trading, both of which employ the Rule of Three. Another important feature of the calculation rules in the *siddhānta*s is a certain level of abstraction of numbers and quantities. Instead of demonstrating examples with concrete numerical values, the rules are described in words. For example:

> In a Rule of Three there are an argument, a result, and a desired quantity: the measures of the first and the last are similar. The desired quantity multiplied by the result and divided by the argument is the result [of the desired quantity]. (Plofker 2007: 422)

A common feature of the *gaṇita* in a *siddhanta* and the stand-alone mathematical works is that they follow the "rule and example" style of exposition, followed by a verification procedure: first a general rule is stated, then an example is given, followed by a demonstration of the rule with a verification of its correctness.

Stand-alone mathematical treatises appeared as early as the eighth century. Rules and examples were composed in verse while solutions were given in prose. Among the problems, calculations included the following typical scenarios:

1. Five merchants jointly purchase a piece of jewelry and pay for it in certain ratios.
2. The total amount of gold is calculated given pieces of gold of various levels of impurity.
3. Taxes are calculated for a variety of garments.
4. Merchants give money for religious offerings according to a progressive arithmetic series.
5. Money is earned following geometric progressions.

In a *siddhānta* of the twelfth century, there are problems of determining interest based on a given amount of borrowed principals, and of determining the value of an ox according to its age based on the assumption that the price is inversely proportional to its age (Plofker 2007: 456). These sorts of problems suggest how mathematics was used by the general population. It should be kept in mind that the audience for written mathematical knowledge in India varied among four specific social classes (*varṇas*) of Hinduism, namely priests, warriors, merchants, and at the lowest, laborers, menial workers, and artisans (Plofker 2007: 477). The composition of works on calculation in India do not seem to have been inspired by any of the affairs of daily life of the population at large, in contrast to the examples found in many contemporaneous Chinese mathematical treatises.

Transmission of mathematics often followed routes of commercial trade and the spread of religion. How certain Indian mathematical and astronomical knowledge found its way to China has been discussed elsewhere in this Volume. Evidence also shows that astronomy and astrology in India were influenced and inspired by earlier Babylonian and Greek traditions. Similarly, the decimal place-value number system and certain trigonometrical calculations were most likely transmitted from India to Islamicate regions of the world. In the eighth century, a *siddhānta* was either translated or adapted into an Arabic work by scholars at the court. Like their Sanskrit counterparts, certain astronomical and astrological texts in Arabic imitated the practice of reciting technical treatises in verse. They also reflect their origins in Indian mathematics by the specific parameters, cosmological models, and computational techniques they adopted, including the use of decimal place-value arithmetic and sine lengths instead of chords in trigonometric calculations in Islamic astronomy.

One encyclopedic work of the thirteenth century was Varāhamihira's *Bṛhatsaṃhitā*, primarily devoted to astronomy but including such diverse topics as combinatorics, comets, eclipses, the significance of planetary conjunctions, astrology, omens and portents, the weather, and other such diverse topics as gems, perfumes, agriculture, architecture, temples, botany, and medicine, among many other topics.

MATHEMATICS IN ISLAMICATE COUNTRIES

Mathematics in medieval Islam developed on foundations influenced by both Greek and Hindu traditions. Numerous Greek works, including Euclid's *Elements*, Apollonius' *Conics*, and Archimedes' *Measurement of the Circle* were translated into Arabic in the ninth and tenth centuries. Judging from available evidence, certain types of mathematics originated with practitioners, including surveyors, builders, artisans in geometric design, government officials, and merchants.

Mathematics in the service of the court included astronomy and spherical trigonometry that was applied to a geocentric cosmographical model with a celestial sphere. The numerical system was sexagesimal for calculations in trigonometry and astronomy, although the decimal system was used in other arithmetic calculations. In plane trigonometry, instead of utilizing the chord extending the angle or arc to carry out calculations, the sine and other trigonometric "functions" were defined as line segments associated with angles in a circle. These definitions are equal to their modern trigonometric-function counterparts multiplied by the radius *r* of the circle.

One major tool in trigonometric calculations were trigonometric tables. To construct a decent table was a considerable feat, requiring in the case of a sine table, for example, knowledge of the sum and difference formulas for sines and cosines proved by Abū al-Wafā' and an accurate calculation of the sine of 1° was another important prerequisite. Al-Kāshī found an approximation to by finding an approximate solution to a third-degree polynomial equation. To utilize the table in calculations, both linear and second-order interpolation procedures were put forth. Al-Bīrūnī in his work *On the Determination of the Coordinates of Cities* introduced procedures to calculate the radius of the earth (which he calculated as 6,339.9 km, only 16.8 kilometers short of the currently accepted distance of 6,356.7 km). Another important property for land surveying using trigonometry is the law of sines for plane triangles proved by Naṣīr al-Dīn al-Ṭūsī.

To determine distances on the celestial sphere, it was necessary to introduce spherical trigonometry. Abū al-Wafā' in his astronomical handbook, *Zīj al-Majisṭī*, presented two major theorems in spherical trigonometry: the law of four quantities and the law of spherical sines. Islamic astronomers and navigators used astrolabes to measure altitudes above the horizon for an observed planet or star. The earliest extant Arabic work on how to use an astrolabe is by 'Alī ibn 'Īsā, who demonstrated five different ways to use the instrument. Another important "application" of spherical trigonometry is to find the *qibla*, the direction of Mecca towards which Muslims must turn five times a day for prayers no matter where they may be. Al-Bīrūnī, in his work on mathematical geography, *The Determination of the Coordinates of Cities*, presented four different methods to find the *qibla*. As an indication of his tolerance towards other religions, he stated that the same techniques could also be useful for Jews to face Jerusalem and for Christians to face east (Berggren 2007: 634–7).

The achievements of medieval Islamic mathematics in algebra are well known. Although symbols representing unknowns had not been invented, the content of algebraic works clearly described in words was equivalent to using an unknown designated as a root or a thing, then describing how its powers, numbers, and the arithmetic operations on "both sides of the equation" should be manipulated in order to reduce the equation to a form from which the value of the unknown in question could be determined. The techniques for solving equations of higher orders (e.g., three or five) arose from the need to solve such equations in trigonometric or geometric calculations. Procedures for finding solutions for simultaneous linear equations, as many as three equations in three unknowns, were also treated. It should also be noted that certain scholars attempted to "geometrize" algebraic results by representing them geometrically. For example, Thābit ibn Qurra used geometry to provide a rigorous proof of the correctness of algebraic solutions for quadratic equations (Berggren 2007: 518).

How did people in Islamicate countries use mathematics in medieval times? Al-Khwārizmī's preface to his *Compendium on Calculation by Completion and Reduction* explained how mathematics was needed "in cases of inheritance, legacies, partition, lawsuits, and trade, and in all their dealing with one another, or where the measuring of lands the digging of canals, geometrical computation, and other objects of various sorts and kinds. . .." (Al-Khwārizmī, quoted from (Berggren 2007: 542)). And just as in mathematics of other cultures, examples demonstrating procedures also presented information on how mathematics was used for practical purposes in astronomy, astrology, engineering, architecture, and everyday business practices. For example, in Al-Kāshī's *Key to Arithmetic*, problems occur involving the value of jewelry in the same box, computation of wages for day laborers, prices of various commodities, inheritance problems, and calculations in geometry, including the surface area and volumes of domes and *muqarnas*. His work is especially notable for its use of decimal fractions and an algorithm to determine the *n*th root of any given number.

What stands out in this list of diverse applications of mathematics are the problems of inheritance, in comparison with model problems found in mathematical works in other cultures. The problems and their solution suggest that Islamic law strictly regulated the shares surviving family members should receive as inheritances, which further complicated matters. In the second half of Al-Khwārizmī's book on algebra, all of which is devoted to problems of inheritance, he includes the following problem: "A woman dies, leaving her husband, son and three daughters, but she also bequeaths to a stranger 1/8 + 1/7 of her estate. Calculate the shares of each" (Berggren 1986: 65). Problems of this kind reflect the importance and the level of complexity of such matters in Islamicate societies generally.

One telling characteristics of Islamic mathematics was its merging of theory and applications. In astronomy, for example, many techniques were derived from concrete needs for specific calculations. In geometry, many scholars addressed the correctness of theorems by providing proofs following the Greek tradition. In the latter part of the tenth century, the mathematician and astronomer Abū Sahl al-Kūhī solved a group of problems related to a tower on an island in the sea solving two unknowns based on a given third. Abū al-Wafā', author of the treatise *On the Geometric Constructions Necessary for the Artisan*, collaborated with craftsmen who frequently cut up squares, rearranging them into other designs. He described how they treated the issue of correctness as follows: "the artisans [have erred] because they lack knowledge of proofs. . . . [For him] correctness (to the artisans) is apparent by what he perceives through his senses and by inspection. He is not concerned with the proofs by means of lines" (Abū al-Wafā', quoted from Berggren 2007: 612). Abū al-Wafā' then went on to describe a particular problem, to construct a square equal in area to three given areas. The geometer proceeded to construct a line from which a square was generated equivalent in area to the sum of the three given squares, but despite the logic of the construction, artisans were not convinced. Instead, their approach was to cut up the three given squares and then rearrange the pieces to form a new square, providing a proof of the given proposition without words—with no need for a geometric construction. This approach was very similar to the ancient Chinese method for establishing the correctness of theorems (like the Pythagorean theorem) related to the computation of areas, as explained in the Introduction to Volume 1 of this series.

A somewhat peculiar example of the intersection of mathematics and a non-mathematical field is a treatise by Aḥmad ibn Mun'im, *Denumerating the Words Such That a Person Cannot Express Himself Without One of Them*. In this case, the field is linguistics. The mathematics treated in this work falls under the category of modern-day combinatorics. What Ibn Mun'im did was to use mathematics applied to linguistic rules to count all the words in Arabic less than 10 letters (Berggren 2007: 660–5). It has been speculated that the purpose of this treatise was purely intellectual. Another Islamic contribution to combinatorics concerns the construction of magic squares. Ibn al-Haytham gave a procedure for constructing magic squares of odd orders. In a treatise attributed to the 10th-century alchemist Jābir ibn Ḥayyān, *The Small Book of Balances*, the magic square and its related numerology were associated with alchemy. Many Chinese mathematical treatises composed after the fifteenth century, when maritime trading with merchants from the West was ongoing, including Islamicate countries, provide numerous interesting examples of magic squares, although drawing any conclusions about specific direct connections is tenuous.

Islamicate contributions to mathematical optics were also significant and had their origins in astronomical theories about the nature of light. Al-Kindi was one of the earliest important optical writers in the Islamic world. In a treatise later translated into Latin as *De radiis stellarum*, al-Kindi wrote "that everything in the world . . . emits rays in every direction, which fill the whole world" (Lindberg 1976: 19). Ibn Sahl, the first Islamicate scholar to write a commentary on Ptolemy's *Optics*, in the course of his mathematical treatment of refraction discovered the equivalent of Snell's law. But it was Ibn al-Haytham (Alhazen) who further developed the subject, believing that vision was the result of rays of light entering the eye from distant objects, thus rejecting the Ptolemaic view that vision was the result of rays emanating *from* the eye. His *Kitāb al Manāẓir* (Book of Optics) was translated into Latin as *De aspectibus* (On Perspective), and was especially influential on the later theories of Roger Bacon and Robert Grosseteste.

Among conclusions both Ibn al-Haytham and Ibn Sīnā reached was that the speed of light must be finite; al-Bīrūnī went a step further and concluded that

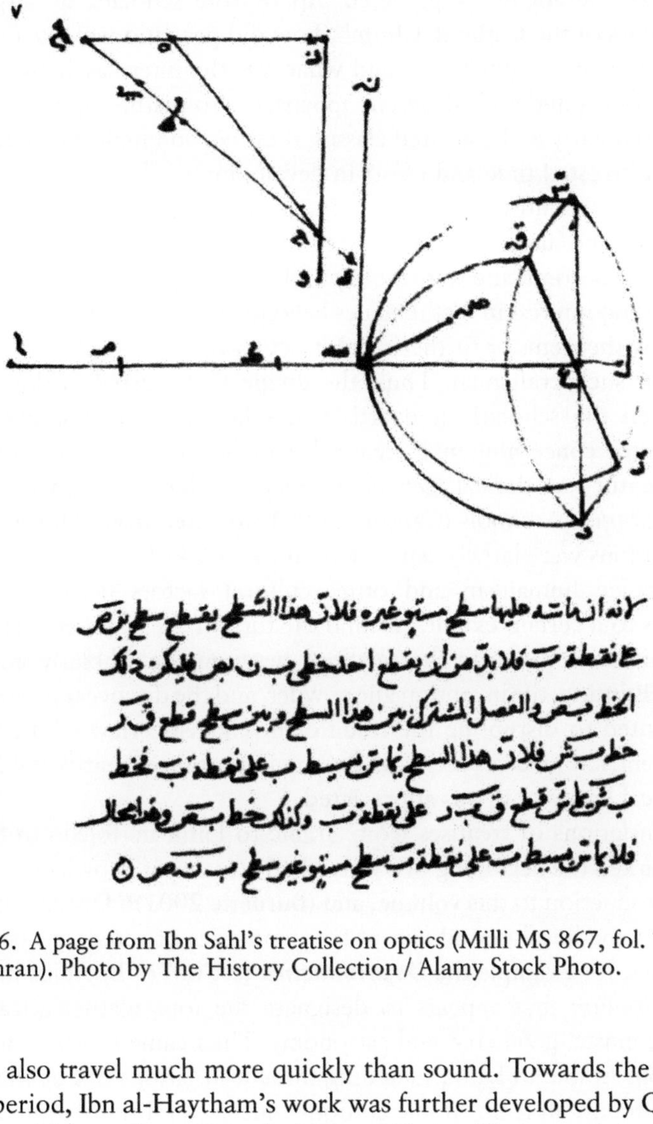

FIGURE 5.6. A page from Ibn Sahl's treatise on optics (Milli MS 867, fol. 7r, Milli Library, Tehran). Photo by The History Collection / Alamy Stock Photo.

light must also travel much more quickly than sound. Towards the end of the medieval period, Ibn al-Haytham's work was further developed by Quṭb al-Dīn al-Shīrāzī and his student, Kamāl al-Dīn al-Fārisī. Their work is notable for having reached an explanation of the optical phenomena of rainbows, a result al-Fārisī included in his *Kitāb Tanqīḥ al-Manāẓir* (Revision of the *Optics*), which in fact was a revision of Ibn al-Haytham's treatise on optics.

MATHEMATICS IN THE LATIN WEST

To describe the culture of mathematics in the Latin West, in the first instance the importance of translations of classics from Arabic versions and later Greek original texts cannot be exaggerated. Apart from scholars, usually clerics in various orders of the Catholic Church, how did practitioners like craftsmen of various sorts apply mathematics and what was the interplay between scholars and practitioners and their different approaches to mathematics?

For the scholarly and educated classes, they had an intellectual interest in the subject and invested time and effort in developing mathematics for the sake of the subject. In contrast, common practitioners were mostly artisans and craftspeople who used mathematics in their everyday work. Mathematical cultures of this second kind were mostly oral, without much advanced knowledge and perhaps no interest in mathematics beyond what was needed for their trade. Proving new theorems or furthering abstract knowledge would not have been a concern for such craftsmen. Thus, the divide that existed in the Latin West was between the scholarly and artisanal, echoing the dichotomy drawn by Moritz Cantor concerning mathematical schools which existed throughout the fifteenth century as clerical (*geistlich*) versus secular or commercial (*weltlich oder kaufmännisch*) schools (Cantor 1900: 166), cited from (Høyrup 2007: 3). These traditions were largely separate until about 1400, when the development of Renaissance humanism and other cultural factors began to blur such distinctions to a certain extent. Around the turn of the sixteenth century, there was a significant change in the relations between the scholarly and artisanal traditions. Printing, the invention gunpowder, and the discovery of the Americas, all contributed to disrupting the traditional scholastic view of the world as a finite geocentric universe. Increasingly, the importance of artisanal knowledge and technical know-how was appreciated.

The translations of treatises from Arabic to Latin in Toledo in the twelfth century marked the beginning of a watershed transfer of knowledge (for details, see the Introduction to this volume, and (Burnette 2001)). Documents are scant for the early medieval period except for summaries of certain parts of classical mathematics, for example the *De arithmetica* (*c.* 500) of Boethius in which the word *quadrivium* first appears to designate the four mathematical sciences: arithmetic, music, geometry, and astronomy. Then came translations of Greek classics from Arabic to Latin, including three translations of Euclid's *Elements* still extant today, all from Arabic editions: one a literal translation, another an abridgement, and the third a reworking or an "editio" (for details, see Clagett 1953). Other sources for activities in geometry include early mediaeval Latin non-mathematical works containing some accounts of geometry and, in the artisanal tradition, handbooks of practical geometry, including surveying. The latter formed a branch of geometrical studies for the most part independent of Euclid (Evans 1976). In contrast to these, the translation projects in twelfth-century Toledo were made in the usual scholarly tradition.

There are few indications of how mathematics was used by those working in the artisanal tradition prior to the thirteenth-century. This is not surprising as the secret technical know-how of a trade was carefully preserved through oral transmission from masters to their apprentices. But thanks to recent studies, more records of how mathematics was used by artisans and engineers during the medieval period are now available. An example of elaborate trading procedures around the end of the thirteenth century demonstrates how nontrivial arithmetic was required to ensure that transactions were equitable among all parties involving, for example, a wool-importing merchant in Florence, a wool supplier in London, a bank in Florence receiving payment from the merchant, and a London agent for the bank who paid the supplier (Biggs 2009). The complexity of the process included currency exchanges, different monetary units, various fees, and possible customs duties.

Another example is the alloying of coinage. Detailed knowledge of how to calculate the proportions of two coinage alloys of different compositions that were needed to produce a third alloy of specified composition, eventually known as the rule of alternate allegation, was a specific requirement in the trade. The calculations were rather involved and included repeated multiplications and divisions of integers as well as of mixed numbers. Translations of Arabic mathematical treatises explained basic arithmetical operations using Indo-Arabic numerals and the decimal place-value system, which greatly simplified calculations rather than using Greek or Roman numerals (Williams 1995).

Expanding trade greatly stimulated commercial mathematics, for which the concept of debt helped to justify negative numbers and their appearance in the solutions to actuarial problems. This was supported by an education system prompted by the rise of the merchant class, namely the so-called abbaco schools. Some suggest this development was a result of the rise of merchant/artisanal culture, others attribute the emergence of the abbaco schools in Italy directly to Fibonacci (see Branson 2013, Høyrup 2023, and Swetz 1987). These were cultural institutions established in northern Italian cities, many of which, in the late 1100s, exiled the nobility and became communes ruled by merchants. The merchants saw the value of training their children in basic grammar and mathematics and provided schools to do so. Such schools are first mentioned late in the thirteenth century. In 1347, a description of Florence mentions the widespread appearance of abbaco schools in the area. Abbaco schools with their commerce-oriented curriculum placed heavy emphasis on practical mathematics. They mostly taught commercial arithmetic to young boys from families of commoners. The schools and the merchant class that supported them were also social and in a certain way political entities. For example, the merchant class created the profession and practice of accounting, which abbaco schools included as part of their curriculum. They played an important role in justifying business as a legitimate social occupation and also in qualifying merchants as people who deserved a significant say in how society was run.

This offers, needless to say, an important insight into worldviews influenced by mathematics in terms at once economic, political, and social.

One particularly significant development of mathematics applied to natural phenomena in the Latin West was further development of the optical theories that were transmitted from Islamicate sources. Robert Grosseteste, an English bishop inspired by the biblical account of the role light played in the creation, used it as the foundation for his explanation of the origins and nature of the physical world. Greatly influenced by Al-Kindī's works in optics, known to him through Latin translations, *De radiis stellarum* and *De aspectibus*, Grosseteste in turn was a major influence on the optical theories of the Franciscan polymath Roger Bacon, whose *Opus Majus* of 1267 included an entire section devoted to optics, which also drew on the prior works of Ibn Sa'īd and Ibn al-Haytham.

Further advances were made by the natural philosopher Vitello (Witelo), whose *Opticae libri decem* (Optics in ten books) focused on individual droplets of water as the optical agent responsible collectively for the rainbow. Vitello's *Perspectiva* also played an influential role in later optical theories of Johannes Kepler. However, the most important of all contributions to the study of optical phenomena in the medieval period were those of the German Dominican Theodoric of Freiburg, who made "the greatest contribution of the medieval age to physical science" (Boyer 1959: 110).

Theodoric, in addition to his mathematical analysis, provided an experimental model by which he could test his theory of how light, by reflection and a double refraction, could account for the phenomena of rainbows. Using a glass sphere of water, he was able to trace the way light entered the glass, was refracted into

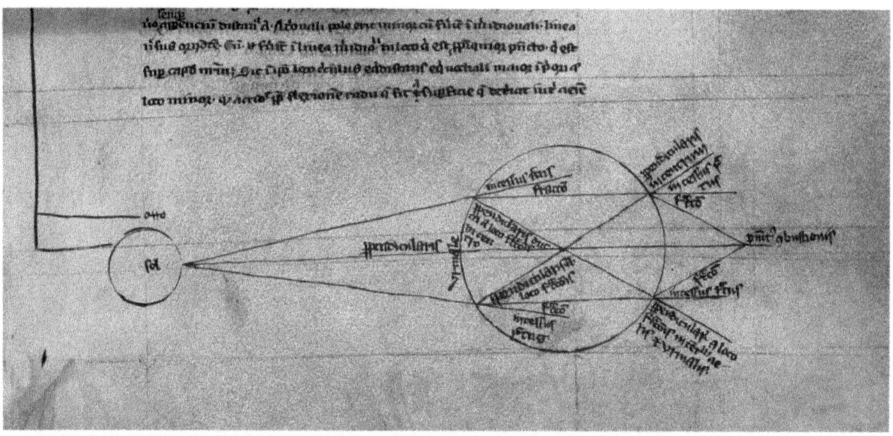

FIGURE 5.7. Roger Bacon, schematic drawing of light refracted through a glass sphere filled with water, from *De multiplicatione specierum*, late-thirteenth-century (MS Royal 7 F. VIII, f. 25, The British Library, London, UK). Courtesy of the British Library, London / Wikimedia Commons.

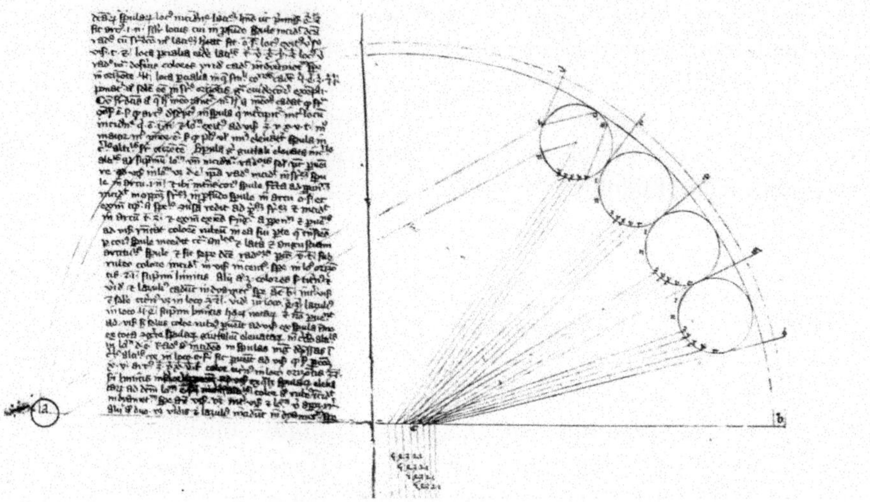

FIGURE 5.8. A page from Theodoric of Freiburg's *De Iride* III,2,5, accounting for the primary rainbow by reflection of light through individual raindrops (Manuscript Basel F. IV.30, folios 33v–34r, Universitätsbibliothek Basel).Courtesy of the University Library Basel, Basel, Switzerland/Wikimedia Commons.

the water, reflected by the glass to another edge where it was again refracted as it left the sphere of water. He went on to explain the secondary rainbow as due to two refractions and two reflections, and by a careful geometric analysis of the path sunlight must follow before reaching the observer, he was able to account not only for the inversion of the colors in the secondary rainbow, but for the weakness of the secondary rainbow due to the second reflection of light through the raindrops that accounted for both the primary and secondary rainbows.

HOW MATHEMATICS WAS TRANSMITTED

We end this discussion of how mathematics was crucial to describing and understanding the medieval world by considering how this knowledge was transmitted and extended between cultures throughout this period. Beginning with China, it always maintained contact with the rest of the world throughout its long history, even if sometimes it remained politically isolated. Multiple Silk Roads between China and the "West" provided connections by overland routes as well as by sea for the exchange of goods and ideas. Moreover, there is evidence to suggest that commercial trade along the Silk Road started as early as the first century CE. NB: by the "West" we here use the Chinese way of referring to numerous regions in the general direction of the west: the dessert terrain of western China, India, the Islamicate regions, and European countries.

For the transmissions of mathematical works, these have been studied and documented although the exact pathways involved in the diffusion of ideas in general are difficult to establish. Here we discuss the known transmissions of mathematics to and from China during the medieval time. The countries or cultures involved are India, the Islamic countries, Korea, and Japan.

Buddhism and Buddhist classics were introduced into China as early as in the first century CE. The terms for the integers, large or small, can be found in many Buddhist classics. Some translated classics during the Sui and Tang Dynasties also contained the twelve signs of the zodiac. The travel records of the Tang dynasty monk Xuanzang, who journeyed to India and brought many Buddhist classics to China, included a brief introduction to the Indian calendar used at the time. In the same century, Indian astronomers of three schools: Kasyapa (*Jiaye*), Kumara (*Jumoluo*), and Gautama (*Qutan*) collaborated with their Chinese counterparts at the imperial astronomical bureau in designing calendars. Among Indian astronomers working in China, the most influential was Qutan Xida, who was named director of the bureau starting in 710 CE. In 718, following a court decree, he compiled the *Astronomico-Astrological Canon of the Kaiyuan Reign*, which contains one of the earliest references in China to a means of representing zero. He also translated certain Indian astronomical treatises into Chinese, such as the *Calendar of Nine Persistent* [*Stars*]. This calendar introduced the measure of dividing the circle into 360 degrees with each degree divided into 60 minutes. Moreover, a table of sine lengths inside a circle of radius 3438 is part of the calendar. The table's range is from 0 to 90 degrees with an increment of 3 degrees and 45 minutes. According to Jean-Claude Martzloff, the *Calendar of Nine Persistent* [*Stars*] is the only one of Indian origin in the *Kaiyuan* astrological canon, which was kept secret and was rediscovered in the seventeenth century hidden in a Buddhist statue. Consequently, its influence on the development of astronomy or trigonometry in China cannot be found in any extant work. Nevertheless, this calendar was apparently based on Varāhamihira's *Pañcasiddhāntikā* and was probably influenced by Brahmagupta's *Khaṇḍakhādyaka* as well (Yabuuti 1979). For further discussions of these matters, see Martzloff (2006: 101–5) and Li and Du (1987: 106–7).

Scientific transmission from Islamicate countries to China was mostly in astronomy, including observational instruments and procedures for carrying out calculations in order to make predictions. Around 1251, a Muslim astronomer, Zhamaluding, arrived in China and in 1267 he presented a number of astronomical instruments to the throne, along with the *Wannian li* (Myriad Years Calendar). Although the identification of this Zhamaluding is uncertain, a number of the instruments he brought with him resembled those used at the time in Maragha, and some scholars have suggested that Zhamaluding was the Persian astronomer Jamāl al-Dīn (Martzloff 2006: 102). It should be noted that during the Yuan and Ming dynasties, many Muslim astronomers worked in the Islamic Astronomical Bureau first headed by Zhamaluding when it was established by order of the emperor in 1271.

Among the books and treatises Zhamaluding brought to China was a version of Euclid's *Elements* in fifteen books, but there is no evidence these were ever translated into Chinese. None have survived, but a record of book titles was recorded in the official dynastic history, and based on a philological analysis, historians believe that these fifteen books were an Arabic version of Euclid's *Elements*. As a final comment on transmitted works, the gelosia method of multiplication (lattice multiplication) can be seen in the 1450 *Comprehensive Collection of Computational Methods in Nine Chapters with Analogous [Problems and Rules]*. This method does not use the abacus and is believed to have been transmitted to China in the fifteenth century, although Martzloff suggests that Chinese mathematicians may have developed the method independently (see Martzloff 2006: 92).

Concerning the dynastic calendars, the most accurate and sophisticated due to the mathematical methods it employed was the Season-Granting Calendar adopted by the Yuan Dynasty (1271–1368) at the behest of its first emperor, Kublai Khan. This calendar was largely the work of Guo Shoujing and Wang Xun, along with others working in the Chinese Astronomical Bureau in Beijing—at the time known as Dadu. Inspired by Jamāl and his astronomical instruments, Guo developed a torquetum that helped to improve the accuracy of the astronomical data used in compiling the Season-Granting Calendar. He also built a gnomon some forty feet in height, which so impressed the later Jesuit astronomer Johann Adam Shall von Bell that he dubbed Guo the "Tycho Brahe of China." Under Kublai Khan's patronage, twenty-seven observatories were built throughout the country in order to gather the most accurate astronomical data possible. The astronomical bureau also utilized both equatorial and ecliptical armillary spheres to determine planetary positions for use in constructing the Chinese calendar and almanacs. For further details, see Sivin (2009).

In addition to better data, Guo Shoujing also used third-order difference interpolation and applications of spherical trigonometry to predict more accurately than previous calendars expected occurrences of solar and lunar eclipses, as well as the determination of intercalary months. The tropical year was calculated to be 365.2425 days and a lunar month to be 29.5306 days—values similar to those of the Gregorian calendar, but that was not promulgated until Pope Gregory XIII did so in 1582.

Was mathematics originating in China ever transmitted to Indian, Islamic, Arabian, or Persian regions? Historians have answered this question in diverse ways. Chinese historians of mathematics generally believe that the Chinese method of double-false position made its way into the Islamicate world from the Chinese procedures of excess and deficiency via India. In addition to the close similarity in the nature of the procedures, there is also the name of the method in Islamic mathematics, "ḥisāb al-khaṭaayn," the "computation with two errors," and its Latin version, "elchatayn," which some argue came from Khitan, an ancient nomadic people located in Mongolia and northern China. Other possible Chinese influences on Islamic mathematics include higher order root extraction (Horner's method) found in al-Kāshī's treatise, simultaneous congruence, and decimal fractions which appear in al-Kāshī's calculation of the circumference of a circle. These supposed influences rely mostly on the similarities found in the practices of different cultures and the fact that such notations and methods as those just mentioned appear in Chinese works much earlier than in Indian or Islamic texts. A claim has been made as well that the positional decimal system itself also originated in China (Lam and Ang 2004).

Meanwhile, kingdoms on the Korean Peninsula interacted closely with dynasties on the Chinese mainland, and Korean culture was heavily influenced by Confucian teachings. In 676, the Silla dynasty united the peninsula and emulated the Tang dynasty in establishing a state college. The official history of the Silla dynasty records the number of mathematics professors and the mathematical texts used in the curriculum. Major Chinese mathematics manuals were adopted, including *Yang Hui's Mathematical Methods* and Zhu Shijie's *Introduction to Mathematical Studies*, among other Chinese works, which were also used in the civil service examinations (Martzloff 2006: 106). Additionally, certain mathematical treatises with titles not known in any extant Chinese documents were also part of the material used for instruction.

The Korean kingdoms had long adopted Chinese calendars. Certain monks during the Silla period went to the Chinese mainland during the Tang dynasty to study calendar making and returned to Korea to help establish the astronomical bureau there. During medieval times, whenever the Chinese emperor promulgated a calendar, copies of the astronomical treatises and canons were printed and sent to provincial governments and tributary states, often including Korea. Kublai Khan, for instance, granted almanacs as many as fifteen times.

In 1278, the three-year-old prince, Chungseon (Chinese name Wang Zhang), who later reigned over Korea from 1308 to 1313, was sent to live at the Yuan court. His mother was a daughter of Kublai Khan and his future wife was a granddaughter. As was customary at the time, the heir apparent of a Korean emperor was sent to the capital of the Mongolian court as a political "hostage" of sorts, and while they were raised in Beijing, they were only allowed to return to Korea after marrying a Mongol. Accompanying the heir apparent to Beijing was a Korean court official, Chi Chengzhi, and for what is said to have been fifty kilograms of gold, he was taught the astronomical state secrets of Yuan astronomy, including details of the Season-Granting Calendar of 1281 (Shi

2021: 363–4). Subsequently, a treatise entitled *Rapid Methods and Tables of the Season-Granting Calendar* was published by Chi Chengzhi's successor, Jiang Bao, in 1343. Two of the six calculation procedures used in this treatise are almost verbatim the same as certain procedures to be found in the Song dynasty mathematical treatise, *A Primer to Computational Learning*, composed in 1299 shortly after the prince arrived in Beijing. It is possible that a copy of this work accompanied the prince or Chi Chengzhi when the prince returned to Korea with his officials. In the fifteenth century, Korea reprinted the Song dynasty mathematical treatise, *Yang Hui's Mathematical Methods*. Such treatises also demonstrate the transmission of mathematics from China into Korea during medieval times.

The story of interactions between China and Japan are similar to those with Korea, especially during the Sui and Tang dynasties when contacts were numerous and well documented. In 600, 607, 608, and 614 (Sui dynasty), Japan sent envoys to China, and did so on nineteen occasions during the Tang dynasty. More than five hundred people, including monk scholars who on average stayed in China for about twenty years, studied astrological astronomy, calendar making, and mathematics, among other subjects.

In 645, Japan emulated the Tang dynasty's administrative structure by establishing a bureau responsible specifically for astronomical observations and calendar making. The curriculum to train future functionaries for calendar-calculation at this bureau included studying the monographs on music and calendars from the Chinese official histories and the mathematical treatise of the *Nine Chapters* in particular. Mathematics education in Japan during the seventh and eighth centuries also looked to China's Tang dynasty as a model. A decree of 718 specified certain details for a Japanese mathematics institute. Six of the nine mathematics canons that constituted the curriculum were among *The Ten Mathematical Classics* of the Tang dynasty.

Moreover, the Japanese civil service examination system also had a channel for students who specialized in mathematics. Records show that the mastery of the *Nine Chapters* was essential. Taking one examination as an example, among nine questions, three came from the *Nine Chapters*. There were two standards for passing—However, answering all nine correctly, or one could pass answering six of the nine questions correctly, but if a student failed any one of the three questions from the *Nine Chapters*, the student was considered to have failed the entire examination (Guo 2010: 327). Graduates who passed their examinations held higher ranks in the Japanese civil service system compared to their counterparts in China.

In the thirteenth century, Japan also compiled its own version of *The Ten Mathematical Classics*. Judging by the titles, six came directly from their Chinese counterpart. As for computational devices, the abacus became well established in Japan by the end of the sixteenth-century, shortly after it had been popularized in China, although with some minor modifications. However, there is no evidence that the Chinese abacus was ever transmitted westward. The closest neighbor where such an instrument is to be found is Russia, but there both the structure of the Russian abacus and the procedures used for carrying out computations bear no similarities to the Chinese abacus (Martzloff 2006: 21).

CHAPTER SIX

Mathematics and Technological Change

Building and Measuring in the Astrolabe Age

JOSEPH W. DAUBEN, CLEMENCY MONTELLE, AND KIM PLOFKER

Worldwide technological change during the more than one thousand years covered in this volume was extensive, and almost any of its manifestations could be considered in one way or another as the product of mathematical reasoning. For the sake of coherence and feasibility, the subjects on which this chapter focuses are divided among a few general categories pertaining to mensuration, construction, and calculation. While many of these subjects have been treated to varying extents in previous chapters, here our focus is on the interaction of material technologies with mathematized views of reality. We begin by exploring technological approaches to the quantification of space and time and the accomplishment of large-scale construction and engineering projects. Finally, we examine in greater detail three specific sorts of devices: astrolabes, magnetic compasses, and computing tools.

Technology (named from ancient Greek *technē,* "art, craft, skill" and *logos,* "word, order, reason, logic") is often superficially equated in the modern world to complicated devices generated by advanced applied science and industrial manufacturing. In this view, "technology" seems to imply a form of material culture that pre-modern people did not even possess. More broadly, however, the term encompasses a range of meanings relating to almost any kind of material or conceptual tools, from hammers and spindles and potter's wheels to more abstract inventions such as alphabetization and tabular grids. In this chapter, the medieval technologies we concentrate upon are ones that explicitly relied upon quantitative reasoning, and sometimes upon advanced mathematical training. Their developers combined ancient methods and devices with new inventions and applications for timekeeping, surveying, and engineering projects, in some cases on a very ambitious scale. Our discussion focuses primarily on the development and use of physical tools and structures, rather than the evolution of purely abstract or textual "tools" like computational methods or layout formats for numerical tables or almanacs.

INSTRUMENTS FOR MEASURING SPACE

Measuring space, whether terrestrial or celestial, has connections to mathematics that are lost in prehistory. As time passed, all such applications of mathematics developed increasingly sophisticated devices that would aid in accurate observations and computations. The most basic tools for measuring space in virtually all parts of the ancient world were simple rulers and measuring tapes or ropes used in land surveying. The earliest references to using ropes or cords to measure land include the *harpedonaptai* or "rope stretchers" surveying terrain in ancient Egypt (see the Introduction to Volume 1). Likewise, in the Bible, rope stretching figures prominently in the Book of Job [38:4–5]: "Where wast thou when I laid the foundations of the earth? . . . Who hath laid the measures thereof, if thou knowest? Or who hath stretched the line upon it?"

In ancient India, the early use of ropes as a means of surveying land appears to be reflected in the ritual geometry texts called *Śulbasūtras* or "Cord-Rules." These works, the earliest of which was likely compiled before the eighth century BCE, described the use of marked cords divided into proportional lengths, together with fixed stakes, to lay out geometrical figures in the construction of baked-brick altars for Vedic rites. Later *Śulbasūtras* also explained the determination of cardinal directions from shadows cast on the circumference of a circle by a vertical gnomon at the circle's center.

The gnomon, set up on horizontal ground leveled with pooled water, remained the chief direction-finding device in medieval India. Its use was also prescribed in triangulations for determining heights and distances, as in the double difference technique described in the Introduction to this volume. To quantify greater terrestrial distances in latitude and longitude, mathematicians understood the use of spherical-astronomy methods involving stellar altitudes or simultaneous observations of lunar eclipses. But for practical purposes, as in most other pre-modern cultures, long distances were estimated in terms of journey duration or other physical metrics (e.g., the linear unit *krośa* or *kos* meaning "cry," supposedly the carrying distance of a human voice).

The ancient Romans had surveying devices that enabled them to engineer such necessities as the great aqueducts that brought water to cities, besides laying out fields, camps, towns, and roads. The *groma*, whose Latin name was derived from Greek "gnomon," was a vertical staff topped with a horizontal four-armed cross enabling the surveyor to lay out straight lines and right angles, especially useful in determining the perpendiculars for intersecting roadways. The less common *dioptra* staff had a disk with a scale of degrees and a rotating pointer instead of the *groma*'s right-angled cross, permitting the measurement of arbitrary angles both horizontally and vertically. The architecture authority Vitruvius also mentions the *chorobates*, a levelling device incorporating a water-level and two plumb-lines, also essential in the engineering of aqueducts; it was a variant of the ancient and ubiquitous A-frame level, which together with the set square helped keep measured surfaces flat and true.

Simplest of all the surveyors' tools was the ancient Greek *schoinion*, a cord marked at regular intervals that could be made from a variety of natural fibers and sometimes even of chain link. Fiber cords that could stretch and shrink due to changing humidity were considered less accurate than the rigid measuring rod or *kalamos* made of reed or wood. These were often of standard lengths, like the *akaina* of 6 ⅔ cubits, which corresponded to the Roman ten-foot rod, the *decempeda* or *pertica*. Various versions of all these fundamental tools continued to be used for land mensuration and building in medieval Latin and Greek contexts.

Islamicate cultures also inherited this mensurational toolkit from the ancient Mediterranean world, as well as Indian techniques such as the so-called "Indian circle" circumscribed about the base of a gnomon to determine cardinal directions from shadows. Accurately determined directions were important in Muslim contexts especially because of the need to compute the local *qibla* or direction of daily prayer facing towards Mecca. A large number of surviving sources in Arabic and other languages describe the medieval use and development of technology for terrestrial measurement. This technology embraced both simple physical manipulatives such as rulers and plumb-lines and sophisticated spherical-geometry tools such as cartographic projections from which the user could read off the local *qibla* direction.

A rare record of a large-scale surveying expedition describes a journey on the Plain of Sinjar in Iraq, sponsored by Caliph al-Ma'mūn in the early ninth century to ascertain the north–south distance corresponding to one degree of latitude or 1/360 of the earth's circumference. Observing teams went south and north from a common starting-point, laying out surveying ropes to measure their progress in linear distance, and using solar altitude observations to measure their change in latitude; it is estimated that the terrestrial circumference thus deduced was accurate to within a percent or so of the true value. Simultaneous observations of lunar eclipses in different localities were also sometimes made by astronomers, to determine the localities' separation in longitude based on the difference in local times. One documented example is the observation of the eclipse of May 24, 997 by Abū al-Wafā' al-Būzjānī in Baghdad and by his colleague al-Bīrūnī in Khwarizm in Central Asia.

In addition to land boundaries, town grids and temples needed to be well situated, often with a particular orientation, usually relative to cardinal directions derived from gnomon shadows. Set squares and other devices go back to the earliest times in China, too, sometimes prominently featured in depictions of deities associated with mathematics, as in a relief carving showing the compass and set square (see Figure 6.1).

In China, the gnomon turned out to be as valuable in measuring distances as it was essential for the measurement of time (see the following section). The method of using the shadow of the gnomon to measure the "height" of the sun, namely its distance above what was assumed to be a flat earth, is recorded in the one of the earliest mathematical documents from ancient China, the *Mathematical Classic of the Zhou Gnomon*, dating to the early Han dynasty, compiled most likely in the first century CE, but including material from even earlier periods of China's history. This work begins with a dialogue between a master mathematician and a student. The master explains how it is possible to measure the height of the sun with only a few simple measurements on land. He does so using the standard units of length at that time, of 1 *chi* = 10 *cun* (the *cun* was the Chinese

FIGURE 6.1. Han dynasty stone relief, showing Fu Xi and Nü Wa, mythological figures associated with creation of the set square, compass, and plumb line; according to legend, the emperor Yu (*c.* twenty-first century BCE) was said to have used these devices in his efforts to control flooding in ancient China, early attempts involving mathematics in construction projects meant to relieve one of the most persistent natural problems that continue to afflict China even today. Photo by The Picture Art Collection / Alamy Stock Photo.

"inch," equal to approximately 2.3 cm or .9 of an inch; the *chi* was the Chinese "foot," equal to approximately 23 cm or about .75 of a foot). The master explains to the student that if two poles both 8 *chi* in height (80 *cun*) are erected such that at noon on the day of the summer solstice, the southernmost gnomon casts no shadow, whereas 16,000 *li* to the north (the *li* being the Chinese "mile"), the gnomon casts a shadow of 16 *cun*, then it will be an easy matter to determine the "height" of the sun by a simple application of the Rule of Three. With the given measurements, the height of the sun x could be calculated as $x{:}16{,}000 = 80{:}16$, i.e. $x = (16{,}000\ li)(80\ cun)/(16\ cun) = 80{,}000\ li$.

The height of the sun might also be computed using a more sophisticated method that did not require measuring the distance from a gnomon in the north to one situated in the far south where the measurement of the distance between the two gnomons might introduce needless inaccuracies. This alternative method is also given in the *Mathematical Classic of the Zhou Gnomon*, and was given further development in 263 CE when Liu Hui, the mathematician best known for his commentary on the *Nine Chapters on the Art of Mathematics*, wrote a "tenth" book that later became a classic text on its own, the *Sea Island Mathematical Manual*. This was the "double difference method" whereby the "height of the sun" could be calculated (but still assuming a flat earth) in terms of the distance between the two gnomons and the difference between their shadows. For discussion of this "double difference method" and derivation of the formula, see Li and Du (1987: 75–7), and (Dauben 2021: 328–37). The same method was used to determine the height of any distant object, be it the height of a watchtower, a mountain on a far-away island, or the depth of a well. These were all problems that could be solved by the double difference method (see Introduction, Figure 0.2).

INSTRUMENTS FOR MEASURING TIME

When measuring the heavens, mathematics was crucial for the accurate determinations of calendars and almanacs, especially the timing of the equinoxes and solstices and the length of the year. With the development of various forms of astral divination, it became necessary also to obtain the best possible predictions of portentous celestial events like solar and lunar eclipses, all of which could have dire consequences for those in power. This is why mathematics associated with the imperial calendar in China, for example, was a closely guarded secret, passed down through generations of computational astronomers associated with the imperial courts.

Throughout the ancient world and even into early modern times, essential tools for keeping time have included the gnomon, often incorporated into sundials by the addition of dial plates of various kinds, and clepsydras or water clocks. These and other ancient devices such as the astrolabe and the armillary sphere were modified into many different forms during the medieval period. All of them involved either some form of celestial observation in order to track the regular flow of astronomical time, or some mechanically created simulation of constant flow. The latter type chiefly used the movement of water (or particles such as sand) through an aperture, but better precision was later made possible by ingenious combinations of gears, escapements, and related devices of the makers of clocks, orreries, and planetaria.

In China, significant advances in technology affected the measurement of time in both the celestial and terrestrial realms. In astronomy, the gnomon played a crucial role, as did advances in the constructions of clockwork mechanisms that resulted in more accurate time keeping. In both of these areas, greater precision took advantage of factors of scale.

The earliest instruments for measuring time in ancient China include gnomons and water clocks. The shadows cast by gnomons and related sundials were useless in bad weather and at night, so in addition to water clocks, gauged burning candles were also used to measure the passage of time.

During the Tang dynasty astronomer Yi Xing made notable use of gnomons when he erected a series of twenty gnomons (actually following a suggestion made by Liu Zhuo in 604) aligned along a meridian running the length of China from the north to what is today Vietnam in the south. These provided data from careful observations that proved useful in his construction of the *Dayan li* (Great Expansion Calendar), which was in effect from 729 to 761. The most important innovation was its understanding of the incommensurable movements of the sun and moon, for which Yi Xing devised a new algorithm by which it took into account, more accurately, the variable speed of the sun's motion along the ecliptic. He also used a method of quadratic interpolation with unequal intervals to correct observations in order to improve his prediction of eclipses.

The most famous of all Chinese gnomons for measuring the sun's shadow and determining details of the calendar was constructed by the Yuan dynasty astronomer and polymath Guo Shoujing, creator of the Season-Granting Calendar. This calendar was promulgated in 1280 at the beginning of the Mogul Yuan dynasty at the behest of Kublai Khan. Guo was able to make this the most precise calendar as yet, more so than any previous calendar, thanks to more accurate observational data he was able to collect with the observatory he built under the patronage of the emperor Kublai Khan, the Gaocheng Observatory at Dengfeng in present-day Henan Province. Built in 1276, this was the largest observatory of its day, with a gnomon measuring 9.46 meters high along with a measuring rule to determine shadow lengths that extended further north of the gnomon by another 31.19 meters (the length of the sun's noon shadow at its greatest extent on the winter solstice reaches almost to the full length of this measure).

FIGURE 6.2. The Gaocheng Observatory at Dengfeng in present day Henan Province, built under patronage of the Yuan dynasty emperor Kublai Khan in 1276. As the world's largest observatory in its day, with a gnomon measuring 9.46 meters, it enabled the astronomer Guo Shoujing to promulgate the *Shoushi li* (Season-Granting Calendar) in 1280, in its day the most precise of any of the Chinese calendars. Photo by Takwing Kwang / Wikimedia Commons.

The calendar Guo Shoujing was able to create using the more accurate data he collected also made use of more sophisticated mathematics that he applied to construction of the calendar involving fourth-degree polynomial equations and the use of cubic interpolations for making calendrical predictions. For the tropical year, he used the most accurate length as yet determined by astronomers anywhere, a length of 365.2425 days. He further improved on the accuracy of his measurements, not only by virtue of the size of the instrument he constructed, but by making the gnomon itself more precise by adding a crossbar at the top to measure the exact position of the sun's shadow cast on a scale suspended in a water trough to insure a uniformly flat surface (the original gnomon is now in the Purple Mountain Observatory in Nanjing, where it was moved from Dengfeng for safekeeping during the War with Japan).

Guo used the gnomon to measure the angle of the sun as it moved along the ecliptic and determine as accurately as possible the exact timing of the solstices and equinoxes. At the winter solstice, the shadow cast by the sun is nearly as long as the total length of the *shuigui* (ruler to measure the sky) in the water trough, which explains the sizeable length of the instrument.

The improved measurements Guo made with this gnomon greatly increased the accuracy of the length of the tropical year, determined to be 365 days 5 hours 49 minutes 20 seconds. Three hundred years later, the Gregorian calendar would obtain virtually the same level of accuracy in the new calendar adopted by the Roman Catholic Church in 1582, which determined that the tropical year resulting from the Earth's revolution around the Sun was 365.2422 days. Guo's Season-Granting Calendar was so accurate that it remained the official calendar of China for 364 years. In fact, centuries later, in 1787, the French astronomer Laplace used some of Guo's measurements to check his calculations on the secular changes of the obliquity of the ecliptic and the eccentricity of the Earth's orbit.

Guo's calendar was adopted with only minor changes by the Ming dynasty as the Great Concordance Calendar and was later carefully studied in Korea and Japan, first as a model adopted in Korea in 1441 and then in Japan in 1684. To ensure that this was the most accurate calendar yet devised, in addition to the gnomon, Guo improved other astronomical instruments to make their measurements of celestial bodies more precise, including the square table, simplified armilla, and a water-powered armillary sphere.

Mathematical timekeeping in medieval India ultimately depended on the observations of astronomers using gnomon shadows or an angle-measuring instrument such as a ring dial or quadrant. Beginning in the late first millennium, Indian astronomers also adopted from the Islamic world other observing devices such as the sine quadrant and the astrolabe. These devices, and instruments in general, were designated by the Sanskrit word *yantra*, "mechanical implement, contrivance." The many variants of complicated luni-solar calendar cycles and astrological systems maintained for ritual, administrative, and social purposes also required astronomical observations obtained with the appropriate *yantras*.

Calibrated water clocks replaced astronomical calculation for timekeeping purposes in everyday circumstances, more conveniently if less accurately. The clocks might be of the outflow kind well known in many parts of the ancient world, or the "sinking-bowl" inflow kind (perhaps invented in ancient India) consisting of a small vessel with a narrow hole, set afloat in a larger vessel of water. The flow of water through the hole eventually filled and sank the bowl after a time interval corresponding to the hole diameter.

Some more esoteric instruments for timekeeping were invented or proposed by Indian astronomers whose texts, along with some surviving exemplars, inform us about their design and use. A device related to the sine quadrant was independently developed by Bhāskara II in the twelfth century, and one for star altitude observations (similar to but pre-dating the European nocturnal) by a fifteenth-century astronomer named Padmanābha. Several authors also recorded descriptions of more speculative "inventions" to model the constant flow of time with no interruptions by means of a purported perpetual-motion mechanism, such as an overbalanced wheel with hollow spokes filled with water and mercury.

Indian astronomical texts might also mention setting up a sighting instrument or a diagram of an eclipse to show the predicted celestial event at the predicted time to a patron or the public. However, no recorded details of official "observatory" institutions or their equipment are known until the early eighteenth century, when Maharajah Jai Singh II of Jaipur built observatories in Delhi, Jaipur, and three other cities in northern India. The immense masonry instruments they housed, which allowed for much more precision in observation than small hand-held devices, were initially modeled on those of the fifteenth-century Samarkand observatory of Ulugh Beg. Several of them, however, including the monumental equinoctial sundial called *Samrāṭ-yantra* depicted in Figure 6.3, were designed or modified by the Jaipur court astronomers or Jai Singh himself.

Timekeeping in the Islamic world required not only the determination of astronomically regulated times and dates, such as the moments for daily prayers or the start of a new month at first sighting of the lunar crescent after new moon, but also the coordination and synchronization of many diverse local calendar traditions in different regions. Portable quadrants, astrolabes, parallactic rulers, and sundials—all in a wide variety of technically specialized or improved modified forms—were the standard observational tools for these purposes, but they were soon supplemented by larger fixed devices constructed for the use of observers in officially established institutions. Systematic programs of astronomical data collection have been attested by the early ninth century in

FIGURE 6.3. Samrat Yantra sundial, Delhi, 1724, constructed under the patronage of Maharajah Jai Singh II (Aquatint, T & W Daniell, 1808). Photo by Science & Society Picture Library / Getty Images.

Baghdad and Damascus, perhaps employing fixed instruments such as a mural quadrant (a large angle-sighting device that had to be mounted on a wall rather than suspended from the hand). Masonry instruments became standard features of specially constructed observatories, most notably the thirteenth-century Maragha observatory in northern Iran and its abovementioned fifteenth-century successor in Samarkand.

The quest for mechanically regulated chronometry that would not require ceaseless celestial observations and calculations continued in many parts of the medieval world. The practical functioning of such devices often reinforced their roles as prestige luxury objects or impressive "wonders," as in the elaborate flow-regulated water clocks, complete with decorative automata, described in two early thirteenth-century Arabic treatises. Descriptions also survive of some Islamic water clocks using self-contained or closed-loop mechanisms. One involved an automatically reset sinking-bowl component whose ultimate inspiration may go back to Indian water clocks. Another used a sealed cylinder divided into compartments partially filled with liquid, whose shifting weight as it poured through holes between the compartments drove the rotation of the cylinder and of a pinion that turned an astrolabe dial. Calibrated candle clocks, with mechanisms powered by the changing weight of the candle as it burned away, were also attested in contemporary treatises. Although Islamic writers on various "ingenious devices" attributed their original inspiration to Hellenistic mechanicians such as Archimedes, it is clear that a great deal of medieval experimentation and development enriched this technological tradition, especially in features such as regulatory mechanisms and gear trains.

Experimentation with non-astronomical timekeeping in the Latin West (apparently producing, for example, the sand- or powder-filled hourglass perhaps in the late first or early second millennium) continued through the High Middle Ages and beyond. As in the case of other mathematical sciences, Latin descriptions and translations of Arabic technical works subsequently propelled further innovations. Foremost among these was the verge or crown-wheel escapement, of unknown origin but probably arising in thirteenth-century Europe, that allowed the regulation of mechanical motion without requiring a continuously flowing substance like water or sand. All-mechanical tower clocks began to replace monumental water clocks in European cathedrals and other public buildings, with many mechanical improvements in the post-medieval period.

Among the earliest achievements in the construction of accurate water-driven, mechanical time-keeping devices in China was the astronomical clockwork devised by the Northern Song dynasty polymath Su Song, as described in detail in Chapter 5. His remarkable water-driven armillary sphere and clocktower would not find its equals in the West until the great astronomical clocks of Prague, Venice, Strasbourg, and elsewhere among the great European astronomical clockworks still in existence. Su Song's device predated the earliest of these by several centuries and involved among other innovations an escapement regulator. For a depiction and earlier discussion of this device, see Figure 5.3. In what follows, we continue the discussion of Su Song's water-driven clock begun in Chapter 5, focusing here on details of the mechanism itself.

One challenge for early water clocks was the problem of providing an adequate force required to operate the mechanism. Su Song solved this problem in part by increasing the scale of the clock itself, which was about 40 feet high, powered by a huge water-wheel that served in effect as a sort of escapement device as buckets on its rim would move into place, then fill with water until the weight was sufficient to move the wheel down a notch, when a new bucket would move into place and begin to fill with water until again, the sufficient weight was reached, moving the wheel down another notch. This moving wheel in turn put a series of other devices into play.

When the Northern Song dynasty fell in 1127, the invading army ransacked the Song capital, disassembled the clock, and moved it to the Jurchen Jin "central" capital, Zhongdu, what is today Beijing, but were unable to reassemble the clock and so it lay in disrepair and was soon forgotten. But the treatise Su Song had written and published in 1094 CE survived, and subsequently this has resulted in a number of models and full-scale reconstructions of the mechanism Su Song had built.

The clockwork included figures that sounded the hours with banging drums, clanging gongs, and ringing bells; figures moving on a rotating wheel would appear behind windows that would open to display notices held on placards. Scale models of the Su Song clocktower reconstructed from Su Song's descriptions may be found in exhibits at the National Science Museum at South Kensington in London and in the National Museum of China in Beijing. Full-scale working replicas of the device may also be found in the Science and Technology Museum in Beijing, the National Museum of Natural Science in Taichung, Taiwan, and in the Gishodo Suwako Watch and Clock Museum in Suwa City, Nagano Prefecture, Japan.

The clocktower itself was a symbol of the Song dynasty's superiority over other earlier achievements of a similar sort elsewhere. In fact, the desire for a more precise instrument to determine the exact time apparently arose first when Su Song, on a diplomatic mission to the Liao dynasty (907–1125 CE) ruled by the nomadic Khitans to the north, who controlled much of what is today's Inner Mongolia, found that the Khitan's calendar was more accurate than the Chinese Song dynasty calendar in use at the time. As Liu Heping surmises: "Zhezong's patronage of Su Song's astronomical clock seems to have been clearly aimed at competing with the Khitan Liao for scientific and national superiority" (Liu 2002: 577). Liu goes on to explain that the clock represented the synergy between the heavenly cosmos and the earthly state, and not only imitated the motions of the heavens, but also served to predict affairs of the state. The names given to individual parts of the great device (see the version of its diagram reproduced in Figure 5.3, and another version in Needham, Wang, and de Solla Price 1986: fig. 23) were explicitly intended to reflect the direct connection between the celestial realm of heavenly order and the terrestrial world of imperial dominion, by identifying for example the upper water reservoir as the "pond of heaven" (*tianci*), the main vertical transmission shaft as the "pillar of heaven" (*tianzhu*), the split-ring meridian circle as the "orbit of heaven" (*tianjing*), the oblique gear wheel as the "wheel of heaven" (*tianlun*), the lower bearing beam as the "platform of heaven" (*tianshu*), and the two supporting stands as the "feet of the earth" (*dizu*)

The connection between the ceaseless motion of the heavens and the continuous flow of water was also a powerful symbol of the synchronous parallel between the macro- and microcosm. As Su Song explained:

> The heavens move without ceasing and so also does water flow [and fall]. Thus if the water is made to pour with perfect evenness then the comparison

of the rotary movements [of the heavens and the machine] will show no discrepancy or contradiction; for the unresting follows the unceasing. (Liu 2002: 578; quoted from Needham, Wang, and de Solla Price 1986: 21)

This so appealed to the Song emperor Zhezong that he went so far as to proudly display images of the clocktower in imperial processions, as depicted in an historical scroll painting of a procession of the emperor in 1053, "Illustration of the Imperial Grand Carriage Procession," original now in the National Museum of Chinese History, Beijing (Liu 2002: 578).

MONUMENTAL ARCHITECTURE: GREAT BUILDINGS, GREAT WALL

One of the most remarkable building projects to occupy China over much of its early history was the effort to protect its borders that eventually expanded over the course of centuries to become the Great Wall of China. This was an undertaking that began with the first stretches of fortifications built as early as the Spring and Autumn period that were later connected during the reign of the first emperor of a unified China, Qin Shi Huang of the Qin dynasty (221–206 BCE), prefiguring the Great Wall. For this project, mathematics was essential to the administrators overseeing this monumental work, who were charged with calculating the amounts of manpower needed, materials to be procured, the time it would take to complete a given part of the project, as well as all of the other essential accounting matters related to any major bureaucratic project. Later, technology contributed to accelerating the pace of work and creating stronger materials from which to build the wall. It may seem mundane, but invention of the wheelbarrow was simply a clever application of the principle of the lever, where the wheel is at the fulcrum and the load is carried midway between the wheel and the force applied from the extremity by pushing the wheelbarrow forward.

Technology also served to improve considerably the resiliency of the wall by strengthening the material used, which in the earlier periods had consisted primarily of hewn stone, rammed earth, or sun-baked bricks. By the Ming dynasty, when the last great push was made to complete and refortify various parts of the Great Wall, kiln-fired bricks further increased its durability.

Numerous examples of mathematical problems related to such construction projects may be found in the writings of mathematicians in China between the Han and Qing dynasties. For example, the fifth chapter of the famous *Nine Chapters on the Art of Mathematics* devoted to "Estimating Workloads"—which the third-century commentary of Liu Hui explains means "in order to deal with workloads, needed manpower, and concrete volumes"—opens with a method for determining equivalencies between excavated earth, solid earth, and loose soil (the relative volumes of which are given proportionally as 4:5:3). The first construction problem then follows, given the dimension of a city wall, to determine its volume. The next calls for calculating the number of workers needed to construct a dike of given dimensions given the workload per worker. Later problems add further complications, like different workloads for handling sand, gravel, water, and stone.

Likewise, Problem 5 in the seventh-century *Continuation of Ancient Mathematics* by Wang Xiaotong concerns the construction of a canal and alongside it, a canal bank. Given the dimensions of the canal, the number of laborers involved in digging out the canal, ramming the earth, and then constructing the bank, given a constant volume of earth handled by each laborer, among other things, one is asked to determine the dimensions of the completed rammed-earth dike (Lim and Wagner 2017: 158–62).

A similar sort of "quantity surveying" for construction projects is attested in the sections of Indian mathematical texts devoted to computational "practices," as mentioned in Chapter 2. They tell us little, however, about the physical technologies used in the construction process. Treatises on the disciplines (*śāstras*) called *vāstu* or *śilpa*, involving the design, layout, and ornamentation of buildings or groups of buildings, invoke gridded ground-plans whose orientations and proportions also have sacred or cosmic significance; some of their principles may date back as far as the ritual geometry of Vedic altar construction in the ancient *Śulbasūtras*.

Geometric designs with numerically harmonious proportions were also a feature of Islamic mosques and other monumental architecture. They ranged from elegantly simple rectangular ground-plans to breathtakingly intricate patterns of decorative tilings and *muqarnas*, a form of elaborate ornate vaulting (Figure 6.4). Originating in the ʿAbbāsid Empire, the typically "honeycomb" or "stalactite vaulting" of *muqarnas* was intended to provide a decorative transition from the walls of a given structure to a domed ceiling. Its complicated designs were also meant to symbolize God's universal creation.

Notable for its mathematical analysis of *muqarnas* is a work by Jamshīd al-Kāshī, whose *Key of Arithmetic* devoted a chapter to "On Measurements" where he describes his reasons for including this subject:

> The specialists merely spoke about this (i.e., measuring for the arch (*faq*) and the vault (*izaj*) and besides that it was not thought necessary. But I present it among the necessities together with the rest, because it is more often required in measuring buildings than in the rest . . . (quoted from Dold-Samplonius 1992: 193)

As Yvonne Dold-Samplonius explains in a detailed study of al-Kāshī's achievements, his interest in the computation of surface areas in particular was

FIGURE 6.4. *Muqarnas*, an elaborate three-dimensional decorative device widely used in Islamic architecture from the mid-tenth century on, involved complicated geometric designs such as the one depicted here from the Nasir al-Mulk mosque (the Pink Mosque), in Shiraz, Iran. Photo by Christophe Boisvieux / Getty Images.

most likely because it was useful for determining the quantity of materials needed for construction, like the number of bricks or amounts of paint or gold. As she concludes:

> Measuring the muqarnas is the last section of the extensive book IV, *On Measurements*, of al-Kāshī's *Key of Arithmetic*. From the triangle to the complicated structure of the muqarnas, and including such bodies as dodecahedra and others, al-Kāshī has now exhausted the subject of measuring. The theme has been rounded off, not just from a formal point of view, but more important, as a set of rules for practical application. (Dold-Samplonius 1992: 225)

Further west, one of the great architectural feats of the early Middle Ages was the construction of Hagia Sophia in Constantinople, when Justinian's brickwork cathedral was dedicated in 537. Arches spanning 100 feet supported an impressive 180-foot-high pendentive dome (so named for its innovative use of inverted spherical triangles to carry the weight of the dome to the supporting corners of a square tower).

Considerably later, European architects and engineers were adept at utilizing numbers in their construction projects, and a wide variety of medieval sources testify to mathematical experts contributing to large scale architectural projects. But none in the Latin West could rival the Gothic cathedrals, of which those at Chartres, Rheims, and Köln (Cologne) are among the most famous. Medieval stonemasonry was also known as the "art of geometry": not the geometry of the universities or Euclid, but the practical geometry of artisans who learned their methods largely through the oral and secret traditions of the guilds. With little more than a ruler and compass, they could create monumental structures and intricate designs that were the hallmarks of the Gothic cathedral.

Stonemasons and builders alike regarded themselves as creating houses of worship in the same way as they believed God had created the world—mathematically. The cathedral was a microcosm reflecting the macrocosm of the universe, and the harmonious proportions of the human element, man, in the vision of the famous Vitruvian man, were encompassed, literally, in the forms of the circle and square, symbolic of heaven and earth.

Construction of the twelfth-century abbey of St. Denis near Paris is generally taken to mark the beginning of the great innovation of the Gothic arch in medieval Western European construction. This carried over to the rebuilding of Canterbury Cathedral in England, and found its greatest expression in the cathedrals built in France at Chartres and in Germany at Köln, reaching its climax in the nearly 1660-foot ceilings in the Cathedral at Beauvais, all made possible by the pointed arch that directed the stress of its roof towards the ground and not laterally, which had limited all previous post-lintel forms of construction.

While we could explore the mathematics embedded in the design of virtually any medieval cathedral, one of the most elaborate mathematical programs may be found in the cathedral of Milan, unique in many ways as the only truly Gothic cathedral to be found in Italy, but also representative of European architecture generally given the many hands that were applied in the design and construction of the edifice itself.

Planning for the Milan cathedral began in 1387 with an overall design based on the equilateral triangle, an idea introduced by one of the German architects involved early in its planning, Annas of Firimburg. Numerous artists and architects contributed over the years to its design and construction, including French architects Jean Mignot and Nicolas de Bonaventure from Paris, the German Hans Parler, and Italians like the mathematician Gabriele Stornalco, the polymath Leonardo da Vinci, and the artist Donate Bramante, who submitted designs for the central cupola. The cathedral was consecrated in 1577, but the façade was still unfinished and in fact, construction of the cathedral was only completed in January of 1965.

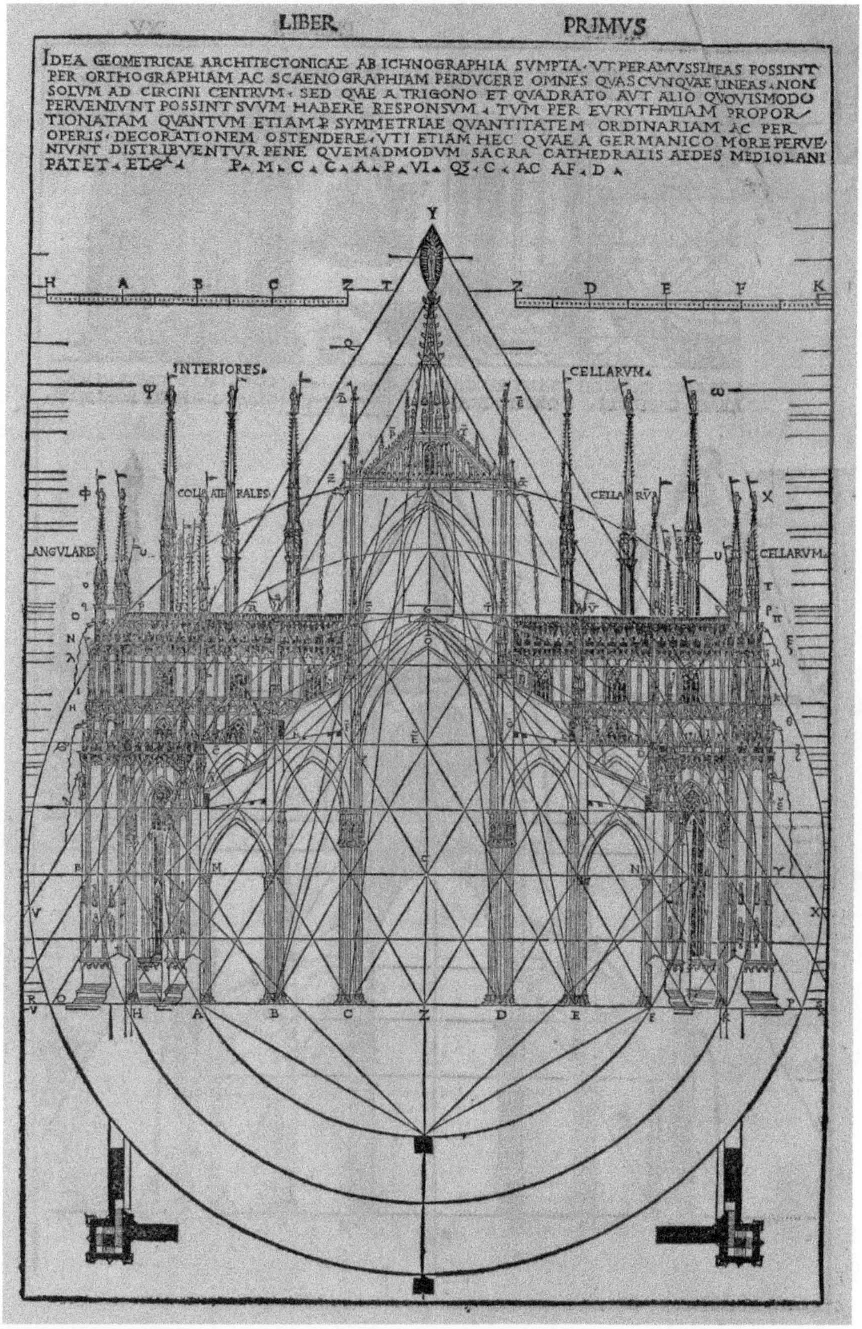

FIGURE 6.5. Geometrical analysis of the construction plan for the Milan cathedral (begun 1387, consecrated 1577). From the Cesariano edition in Latin of Vitruvius' *De Architectura* (Cesariano 1521: XV verso). Image courtesy of the Getty Research Institute and HathiTrust.

Stornalco's original geometric plan for the cathedral began from a regular hexagon inscribed in a circle, wherein the inscribed equilateral triangle determined the height of the nave; the diagonal of the square inscribed in the hexagon determined the height of the imposts of the outer aisle vaults. The Milan cathedral with its complex mathematical proportions became something of an international project, one to which many would contribute and study:

> the architects employed on the cathedral came from many different places, Lombardy but also from Venice and the Ticino, and from the Rhineland, Bavaria, Paris, Normandy, the Low Countries and even Hungary, and it was precisely this that made the enterprise a centre of architectural research and international exchange. The sheer fame of such an ambitious project fascinated master-masons and artisans from all over Europe and it is easy to imagine that the Deputati wanted the very best on offer. (Sebregondi and Schofield 2016: 83)

That the practical mathematics of the stonemasons was of such widespread interest is reflected in the rapidity with which construction handbooks, some of the first printed mathematical works, appeared from the presses beginning in the late fifteenth century: in particular, ones by Mathes Roriczer, Hans Shmuttermayer, and Lorenz Lechler.

OTHER ENGINEERING TECHNOLOGIES

The greatest of all engineering projects in China, demanding much more in technological ingenuity and engineering innovation than the Great Wall, was construction of the Chinese Grand Canal. When completed, this provided both merchants and military alike with a continuous waterway stretching for over a thousand kilometers, from Hangzhou in the south to Beijing in the north. Given that the major rivers of China generally flow from west to east, the challenge of the grand canal was to link the prosperous agricultural south with the northern parts of the country. The idea for such a canal can be traced back to early canals built during the Spring and Autumn Period, but its earliest form was not achieved until after the collapse of the Qin and Han dynasties, followed by some 400 years of civil strife and fragmentation of the country until it was again unified during the Sui, Tang, and Song dynasties.

The Tang and Song dynasties witnessed the second historical wave of canal construction and is regarded as an outstanding period of innovation. The combination of warehousing and waterways supported a thriving economy, all thanks to a key innovation in canal construction, namely that of the chamber or pound lock system, invented in 983 by the provincial transportation commissioner, Qiao Weiyue. Qiao built the first recorded chamber lock on the Grand Canal, basically a section of the canal that is enclosed with watertight gates at both ends so that water can flow in or out, thereby either raising or lowering the water level as needed to transport vessels form one level of the canal to another. The value of this invention is exemplified by the complex gates regulating the Grand Canal where it had to traverse the confluence of the Yangtze and Huaihe Rivers, all reflecting hydraulic engineering at its best. The total length of the Grand Canal was 1,700 km (somewhat more than 1,000 miles). Along the way, lakes and dams provided all of the water needed to maintain and regulate the system.

It was during the Sui dynasty that the full extent of the canal reached as far to the north as Beijing (completed in 609), a crucial connection to facilitate transportation of grain and supplying the army garrisoned in the north to protect China from marauding Mongol invaders. During the Ming dynasty the canal was deepened and new locks constructed to further enhance its role as China's lifeline in the movement of goods and grains from south to north.

Engineering projects in the Islamic world included the construction of underground aqueducts called *qanats*, irrigation systems and water-wheels, mills, dams, bridges, and engines of war, as well as the mining and forging of metals. The mathematical tools they relied on ranged from the ancient and elementary, such as flood gauges and water-raising Archimedean screws, to the theoretically advanced, such as trigonometric shadow squares used for surveying the layout of *qanats*. Although the quantitative tasks of laborers were deliberately kept as simple and routine as possible—reading measurements off scales, adding successive lengths, and so forth—engineers and administrators in charge of projects often had complicated problems to solve.

The relations between Islamicate engineering technologies and those of contemporary South Asia and Latin Europe remain incompletely understood, largely due to the dearth of textual and material evidence concerning them. Quantity surveying problems in Indian mathematics texts attest to early medieval familiarity with construction projects involving, for example, water-tank systems, excavations, and walls. In both India and Europe, some Islamic designs for water-wheels and other machines were adopted in the late first or early second millennium. The difficulty of pinpointing such transmissions is compounded by the fact that the transmitted technologies were accompanied by, and integrated into, active local traditions of technological development. These indigenous traditions seem to have produced, e.g., the origin of worm-gear rollers in early medieval Indian cotton ginning and sugar milling, and the abovementioned development of the mechanical escapement in thirteenth-century European horology.

NOTABLE OBSERVATIONAL AND COMPUTATIONAL INSTRUMENTS

Astrolabes

The planispheric astrolabe was the most mathematically sophisticated and arguably the most iconic technological device of the Middle Ages, functioning in medieval imagery as a talisman of esoteric knowledge and computational power. It originated in Hellenistic geometers' efforts to accurately represent the surface of a sphere on a plane, culminating in what became a standard stereographic projection with the north celestial pole at the center of the device and various great circles, including the celestial equator, arranged concentrically about it. The ecliptic with its zodiacal constellations was shown as a graduated eccentric circle within a network (called the "rete" or "spider") of fixed-star positions. The horizon and zenith of the user's local latitude were similarly projected onto a separate plate. When the latticework of the rete bearing the projection of the heavens was laid over the solid plate bearing the projection of the earth, it could be rotated to simulate the apparent diurnal turning of the sky. The user could then read off from the instrument's markings the time and direction of some celestial event, such as the rising or setting of a particular star.

By the medieval period the plane astrolabe also functioned as an observing device, thanks to the addition of a rotating pointer and 360-degree circular scale on its back. The astrolabe user thus could directly measure the altitude of any sightable object, celestial or terrestrial, without having recourse to a separate protractor-type instrument such as a dioptra or parallactic ruler. Islamic astrolabe-makers often engraved "shadow squares," marked with scales of shadow lengths for a gnomon of a given height observed at a given angle, on the backs of their instruments to facilitate surveyors' computations of heights and distances (see, for example, Figure 0.5).

The computational versatility and power of the standard astrolabe design made it applicable to innumerable problems in astronomy, astrology, and geography, as well as ordinary surveying: it could be used not only to observe celestial bodies' altitudes but to determine from them the bodies' coordinates and the corresponding time of day or night, and many other consequences. Indeed, the tenth-century Persian astronomer ʿAbd Raḥmān al-Ṣūfī ascribed to the astrolabe more than a thousand uses. These functions were enhanced by an immense diversity of innovations and variations introduced over the centuries by Islamicate mathematicians, including multiple sets of markings to accommodate multiple calendars or times of special significance, combinations of different projections within the same instrument, and even a spherical form which represented the movement of the heavens around the earth using no projection at all. Astrolabes might even be combined with geared mechanisms or water clocks in quasi-automated displays of heavenly motions.

Perhaps the most notable, although not the most intuitive or popular, of all such improvements were the mathematical modifications of planar projection formats to render the astrolabe "universal." That is, because the markings on the plate of a standard astrolabe represented only one particular terrestrial latitude, additional plates (which could be stored within the astrolabe's base or *mater*) were required in order for it to be usable at other latitudes. This limitation was overcome by "universal" plate projections devised by, among others, Andalusian astronomer Ibrāhīm ibn Yaḥyā al-Zarqālī ("Azarquiel" in the Latin world) in the eleventh century and Shihāb al-Dīn al-Sarrāj in the fourteenth. The latter invention was especially ingenious in its incorporation of five distinct techniques for manipulating the moving parts to solve astronomical problems at any desired latitude. As David King has written: "I consider Ibn al-Sarrāj's universal astrolabe, which is universal from five different aspects, to be the most sophisticated astrolabe from the Near East and Europe in the entire medieval and Renaissance period" (King 1983: 545).

The ordinary latitude-specific astrolabe nevertheless remained the chief workhorse of medieval astronomical technology. It was enthusiastically adopted beginning in the early second millennium both in India, under the eulogistic Sanskrit name *yantra-rāja* "king of instruments," and in Europe, originally via instruments imported from Muslim Spain and engraved with Latin translations of their Arabic markings. Related devices such as various types of quadrants, sundials, and equatoria also evolved and diffused in these linked astronomical traditions. The so-called mariner's astrolabe discarded the instrument's celestial projections and rete but retained the pointer and degree scale on the back, for measuring star altitudes for navigation.

The Magnetic Compass

The principle of the magnetic compass was first discovered somewhere in the ancient Near East when it was observed that iron oxide minerals naturally aligned on lines pointing north-south (i.e., a lodestone, a stone "pointing the way," from the Old English word "lode" meaning "way" or "journey"). The earliest references to such magnetized stones date back, according to Hellenistic Greek accounts, to the Ionian philosopher Thales of Miletus. The English word "magnet" from the Greek name meaning "Magnesian stone" may echo early associations of these stones with one of the ancient Mediterranean localities called Magnesia.

In China the magnetic compass was devised most likely sometime during the late third-century BCE Qin dynasty, when the same discovery was made, that an iron oxide mineral pointed to the south. This had a particularly auspicious significance in China where it was important to orient major structures on a north-south axis in conformity with the terms of Chinese geomancy, or *feng shui*. This is an occult term meaning literally "wind and water," referring to

FIGURE 6.6. Model of an ancient Chinese compass. Getty Images Stock Photo.

local geographic variables that were regarded as being of primary importance in determining where to locate a house, a temple, a village, or an entire city. The schools of *feng shui* varied throughout China depending upon local topographical conditions. In the south, well supplied with water but with a more undulating terrain, topography was important; in the drier, flatter north, the "compass" school of *feng shui* predominated, and every effort was made to orient any significant construction so that it would face in a southerly direction.

The significance of the "south" in traditional Chinese thought has long been associated with the sun, growth, and prosperity, whereas the character for "north," 北 *bei*, evolved from an ideogram representing two people standing back to back, and may be associated with "back," "harsh," "disagreeable," as well as darkness, the opposite of light and prosperity. Thus, it comes as no surprise that in aligning buildings, according to optimal *feng shui* practice, the backs of buildings should face northwards, whereas the fronts, entrance doors, and main gates should, ideally, face southwards, to receive the full light and warmth of the sun. North and south, associated with darkness versus light, also have correspondences with *yin* (shadow) and *yang* (light), the fundamental principles of the *Book of Changes*. Likewise, there is a long tradition in China of the south-pointing chariot, a vehicle with a clever gearing mechanism such that no matter in what direction the chariot might move, a figure atop the device with an outstretched hand would always point to the south.

Carefully planned cities faced south, government buildings and palaces whenever possible would face south, the emperor sitting on his throne, facing south, would have his back to the north. Accordingly, ancient Chinese rulers sat on south-facing thrones in south-facing palaces, reflecting an old Chinese saying, "the king faces south." Indeed, Czech archaeologists who undertook careful surveys of such important sites as the Forbidden City as well as the much smaller Temple of Heaven and others for the Earth, Sun, and Moon in Beijing, as well as Kublai Khan's summer capital of Xanadu (Shangdu), were all sited using the north-south principles of *feng shui*.

The magnetic compass in both water-floated and "dry" versions entered the Islamic and European technological spheres around 1200, for navigational as well as astronomical purposes (although magnetic deviation made its fulfillment of the latter function only approximate). A magnetized metal pointer and a circular rim marked in degrees, as shown in Figure 6.7, enabled the medieval compass user to read off the desired direction more or less as modern compass users do. It was valued particularly in the Muslim world as a *qibla*-finding device: with a compass and a precomputed list of *qibla* directions corresponding to various localities, a Muslim in any of those localities could easily determine how to face towards Mecca with no astronomical calculations required.

Counting Tables, Abacuses, Mechanical Computers

The earliest devices in ancient China to aid in basic arithmetic computations were counting surfaces where objects representing units and, depending upon their positions, usually powers of ten, could be placed and then manipulated to add, subtract, multiply, and divide. The Salamis Table from ancient Greece (discussed in Volume 1) is a counting surface that can be considered as an early form of the (Western) abacus, using loose counters rather than the fixed manipulation of beads on a rod. Some early Indian sources allude to pits and counters employed in place-value calculation; it is speculated that later texts denoting unknown quantities by a word meaning "ball" or "bead" and by color names may even indicate the use of colored counters for unknowns in manipulative algebra procedures. However, the primary media of mathematical computation in medieval India and the Islamic world were blank writing surfaces such as dust-boards, slates, palm-leaves, and paper, rather than devices pre-structured for number work.

In China the use of a counting board or similar surface for carrying out computations using counting rods to represent units developed into a base-ten arithmetic with which calculations could be carried out quickly and accurately. As time passed, these surfaces were adapted to accommodate the computation of square and cube roots, and even permitted the solution of polynomial equations and sets of simultaneous equations, with designated areas on the

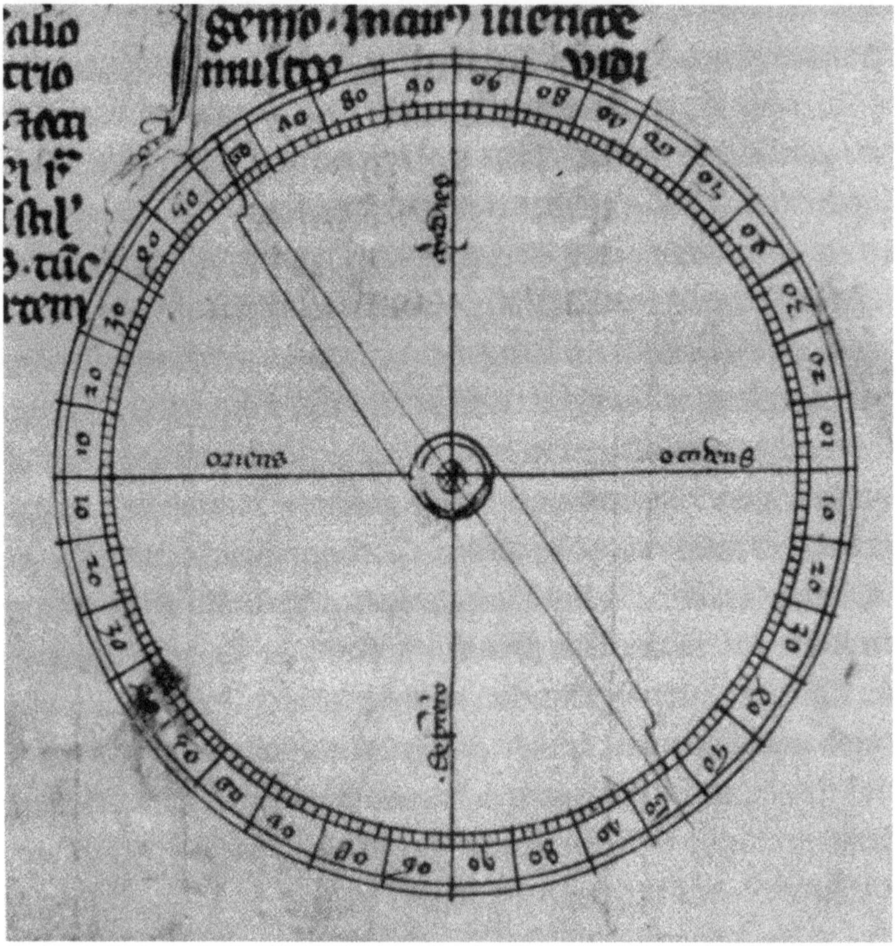

FIGURE 6.7. Drawing of a compass from a fourteenth-century copy of Petrus Perigrinus, *Epistola de magnete* (MS. Ashmole 1522, fol. 186r). Courtesy of the Bodleian Library, University of Oxford, Oxford, UK / Wikimedia Commons.

counting board used to represent unknowns and the various powers to which they might be raised.

In fact, as early as the *Nine Chapters on the Art of Mathematics*, matrix methods had been devised for the solution of simultaneous linear equations in as many as six unknowns. The crucial point here is that the counting board, by designating certain positions to represent each of the unknowns, allowed the mathematician to then manipulate the coefficients exactly as if they were algebraic equivalents. By the thirteenth century, this sort of Chinese algebra had advanced to its highest state, allowing the representation of simultaneous equations in as many as four unknowns, each of the unknowns and their combinations occupying a particular region of the counting board. Since there were only four regions available, top, bottom, left, and right, this method was limited to four unknowns, but it allowed mathematicians like Zhu Shijie to perfect the *tianyuan shu* (method of the heavenly element), the heavenly element being the algebraic unknown (for an example, see Chapter 7, Figure 7.10).

In his book *Jade Mirror of the Four Unknowns*, Zhu Shijie presents 288 problems, and explains how to reduce a problem given in terms of polynomial equations up to the 14th power, in as many as four unknowns (instead of x, y, z, and d, he used the symbols 天 *tian* (heaven), 地 *di* (earth), 人 *ren* (man), and 物 *wu* (matter)), and how these could be successively reduced to a single polynomial equation in one unknown that could then be solved by a method devised by another Song dynasty mathematician, Qin Jiushao. In his fundamental work, *Mathematical Treatise in Nine Sections* (1247), Qin applied a method equivalent to Horner's method for solving polynomial equations. While the advanced nature of these achievements is noteworthy, they were unfortunately short lived. The utility of the heavenly element method depended upon the geometrical disposition of counting rods in locations on the board that represented different unknowns and their various combinations and powers.

This was something the Chinese bead abacus or *suanpan* could not readily accommodate, being a device meant primarily for facilitating arithmetic operations, although methods were devised to go beyond multiplication and division to allow calculation of square and even cube roots. But handling complex problems involving multiple unknowns, allowing solutions of higher degree polynomials and formulating sets of simultaneous equations were beyond the capacity of the abacus. Inevitably, as the abacus came to be widely adopted in commercial settings in particular, the methods of the counting board were eventually forgotten.

By the fifteenth century, Ming dynasty mathematicians had little to no understanding of the methods of the celestial element or of the four unknowns. This was in part due to the cumbersome nature of carrying out even basic operations using counting rods, despite the flexibility they offered on the counting board in accommodating polynomial equations and the solution of simultaneous equations.

In 1592, Cheng Dawei published his *Systematic Treatise on Arithmetic*, where use of the bead abacus was explained and how it could be used to solve a wide range of practical problems. This is said to have represented the end of any interest in counting rod arithmetic, which was replaced by the widespread acceptance of the abacus throughout China. As Li Yan and Du Shiran summarize the situation at the end of this period in the history of Chinese mathematics:

> The completion of the *Systematic Treatise on Arithmetic* and its wide distribution signal the completion of the evolution from counting rod to

abacus calculation. From that time on the abacus became the main means of calculating, the counting rod calculations of ancient times were gradually forgotten and ultimately lost. Thereafter ordinary people only knew abacus calculations and did not know of the existence of counting rod calculations. (Li and Du 1987: 188–9)

The earliest source to which some have pointed for evidence of the abacus in China is a work that may have originated in the Han dynasty, but the earliest written reference to an abacus-like instrument appears in an obscure work, the *Shushu jiyi*, *Notes on the Traditions of Arithmo-Numerological Processes,* also translated as *Memoir on Some Traditions of Mathematical Art*, which survives in an edition with a sixth-century CE commentary by Zhen Luan, in which instruments with beads are described, but no computations or specific uses are mentioned. It is more likely that the abacus first appeared as a means of facilitating mathematical computations sometime during the Yuan Dynasty beginning in the early thirteenth century. By the time of the establishment of the Ming Dynasty in 1368 the instrument was clearly in use, and the first manuals began to appear then explaining how to use the abacus, with explanations given for shortcuts and various ways of carrying out complicated computations. In Japan, the earliest surviving printed mathematical work was the *Book on Division* published in 1622, written by Mōri Shigeyoshi, also known as Mōri Kambei, a Samurai and later provincial governor as well as a master of the abacus or *soroban* (Horiuchi 1994: 30).

The medieval Greek and Latin West retained the counting-board abacus or “Pythagorean table” of the ancient Mediterranean world, where loose counters were combined and rearranged in columns representing successive powers of ten or money denominations. A modified form attributed to the tenth-century Gerbert of Aurillac used counters individually marked with Indo-Arabic decimal digits rather than in groups of undifferentiated units. The word “exchequer” signifying government accountancy is derived from a device similar to the counting board, namely a gridded surface or checkered cloth on which revenue officials calculated amounts due and received. For discussion of the late medieval *Red Book* of the Irish Exchequer, and a figure deconstructing the original illustration of the checkered cloth of the exchequer as depicted in the original book, see Ferguson (1854: esp. the diagram between pp. 46 and 47).

As Indo-Arabic numerals were introduced into Europe along with decimal arithmetic, calculating with written numbers gained in popularity. The *Margarita Philosophica* published in 1508 symbolizes this shift with an illustration of an imagined contest, presided over by the “Spirit of Arithmetic,” between the figure of “Pytagoras” using the old method of the counting board and “Boetius” using the new written decimal arithmetic (see Figure 1.4 in Volume 3). Thus began a steady decline in the use of counting boards and by 1600 they had disappeared almost entirely.

More abstruse calculating tools were available for the needs of astronomers and astrologers, particularly in operating with trigonometric relations. Besides the innumerable written tables of mathematical and astronomical function values compiled in almost every language in which literate astronomy was practiced, graphical diagrams were engraved on angle-finding instruments such as the sine quadrant (or more generally, trigonometric quadrant). On the quadrant's surface, a quarter-circle grid bounded by graduated scales allowed the user to directly read the value of, e.g., the sine of an angle from the point on the sine scale indicated by a cord or pointer set to that angle. Additional curves marked on the grid determined the solar declination or prescribed prayer time or some other quantity corresponding to an observed altitude angle. A "horary quadrant" was marked with curved hour lines similarly calibrated for solar or stellar altitude. Such trigonometric graphs could be engraved instead on the back of an ordinary astrolabe, along with the shadow squares mentioned in the foregoing discussion of surveyor's tools, thereby further extending the marvelous functionality of this medieval "personal computer."

CHAPTER SEVEN

Representing Mathematics

Formats—Disciplines—Practitioners

SONJA BRENTJES AND NATHAN SIDOLI

The history of the term "mathematician" is not well researched for Eurasian and African societies in the period covered by this volume. Mathematical activities on the other hand have received much attention. This chapter has been compiled on the basis of this kind of research. It does not aim, however, to provide a comprehensive survey on all forms of mathematical activities. The chapter will rather focus on a set of basic formats that were used in different mathematical sciences, such as diagrams, and a selection of specific mathematical practices, such as sketches or visualizations of arithmetic procedures. A final part of this chapter will be devoted to the intersection between mathematical sciences and the arts as they occur, for instance, in visual representations in the astral sciences or in divinatory practices. Symbols of mathematical disciplines or mathematical concepts occur in only some of the societies under consideration and individual portrayals of practitioners of mathematical knowledge are rather the exception in this period. The space available for this chapter does not allow us, however, to include in our discussion mathematical and astronomical instruments or representation in mathematical tables.

MATERIAL CONDITIONS FOR THE PRODUCTION OF TEXTS

Structuring mathematical texts through various devices such as titles, tables of contents, headlines, subdivisions, foliations, formulas, diagrams, and other organizational means seems to be an ordinary representational practice from the viewpoint of modern writing, printing, and reading practices. But this was not always the case. In our period, the overwhelming majority of mathematical treatises were handwritten on various materials, such as parchment or vellum, papyrus, paper, birch bark, and palm or banana leaves. The exception to this rule is found in East Asia where wood-block printing was introduced at the end of the sixth century. East Asian print culture and West Asian manuscript culture were materially linked through their use of different sorts of paper. Most treatises meant for the practical use of experts, or in teaching, were written or printed on cheaper types of paper, while expensive, gilded, and patterned papers served for works produced for courts, including mathematical texts. Parchments and vellums were used in the Eastern Roman Empire as well as Western Christian Europe for written work in all intellectual fields until about the fifteenth century, including mathematical works. Around the middle of this century, paper finally replaced those two media as the main material carrier of scholarly productions at universities, scribal workshops, and private homes in Western Europe, while the Eastern Roman Empire was conquered by the Ottoman dynasty. In North Africa, parchment or vellum served as the primary writing material, in particular for sacred books, until about the thirteenth century and paper seems to have entered into book production after 1100, when the first paper mill is recorded for Fez. In the Islamic parts of the Iberian Peninsula ("al-Andalus"), paper seems to have made its entrance some decades later and soon acquired a special reputation for its quality. Birch bark along with palm and or banana leaves became standard material for text production in South, and possibly also Southeast, Asia in the first millennium BCE. While birch bark was mostly used in northern South Asia, palm or banana leaves dominated in the South as well as the Southeast. Paper spread there centuries later; it was the preferred medium of Muslim scholarly and administrative cultures in South and Southeast Asia since the establishment of Turkic dynasties and the conversion of local rulers soon after the tenth century.

In addition to those dominant material carriers of mathematical knowledge in post-antiquity, metals, in particular copper and brass, silk, bamboo, and stone were used in the forms of plates, dishes, temple walls or rocks for royal and other institutional inscriptions of various lengths, containing calendar information, astronomical data, astrological interpretations, and, occasionally, even calculations or drawings.

Those different material carriers of mathematical texts shaped substantially our current knowledge about representational practices in the mathematical sciences, and they will continue to shape any knowledge that we can possibly acquire. The majority of extant medieval manuscripts contain copies of texts that were produced decades, or even centuries, earlier than our current witnesses, although in the case of medieval mathematical writings, we have a few holographs and also a few copies that were directly based on a version of the text made by the author. While a few mathematical texts written on some of the listed materials are preserved from antiquity, most of them were copied in the medieval period. Greek and Latin mathematical texts on parchment are preserved in fragmented palimpsests from the seventh century, and as full texts from the ninth. They were produced, for example, in Syria (London, British Library, Add 17211, Byzantium (Euclid's *Elements*, Oxford, Bodleian Library, D'Orville 301), or France (extracts from Macrobius and Capella; Oxford, Bodleian Library, Laud Lat. 188). Arabic mathematical texts on paper are extant from the middle of the tenth century (Euclid's *Elements*, Tehran, Malik Museum, 3586). The oldest extant mathematical book on birch bark is a medieval copy (apparently eighth to twelfth century; see Figure 3.2) from South Asia of an older work on arithmetic and some geometry (Oxford, Bodleian Library, Bakhshālī). Mathematical texts written on palm or banana leaves are rarely extant before the seventeenth century, although among them copies of much older texts are found, as is the case for the other material carriers. The reasons for these differences in preservation are manifold and range from natural conditions to cultural customs, such as the creation of institutions like libraries or schools, inheritance practices such as donations of manuscripts to religious centers, or the sale of private libraries on the book markets.

A further material condition that shaped forms, durability, or quality of representations of mathematical knowledge or symbols was provided by the writing instruments and inks. Writing and drawing on parchment or vellum was done with a quill. Palm or banana leaves were scraped with a metallic stylus. Quills, reeds, or brushes served in different parts of Eurasia and Africa for writing and drawing on paper. Depending on the qualifications and goals of the manuscript producers, both the writing and the drawing could be executed carefully and in high quality or sloppily. In manuscripts produced in West Asia, North Africa and Europe copyists quite often abstained from copying parts or all of the drawings. In South and Southeast Asian manuscripts, geometrical, astronomical, or astrological diagrams appear significantly less often as regular parts of treatises. Tables, in contrast, are widespread across all those material cultures of manuscript production. Printed Chinese or Japanese books, on the other hand, often contain highly elaborate drawings of mathematical, astronomical, or astrological content.

STRUCTURING MATHEMATICAL TEXTS

The main elements used to structure mathematical texts that developed over time were titles, author names, tables of contents, foliations, headings, coloring, placing markers in the margins pointing to a special part in the text, reserving spaces for drawings, tables, numerical exercises, or other kinds of visual representations of mathematical content, and layout formats such as dividing a text into two or more columns or using different writing directions, types and colors of script (calligraphy), and script sizes. Further elements that guided readers of mathematical texts to important information were references to authors and book titles, numbers of, or descriptive references to, definitions, postulates, axioms, or propositions, or verbal markers that indicated the content of a previous proposition, the beginning of a new part of an explanation or demonstration, or the end of an exposition.

A form shared with many other types of texts was the colophon indicating the end of a treatise, which could contain various statements. The shortest types of colophons merely declared that the book had been completed, sometimes followed by a religious invocation. Longer colophons added a date in one or more calendrical systems and perhaps the name of the scribe. In Western Europe and North Africa, Islamic, Julian, or Hebrew dates could appear together. From Egypt to Anatolia, Iran, and Central Asia, Seleucid, Roman, Sasanian (Yazdigird), and lunar or luni-solar Islamic, dates might be combined. In South Asia various forms of Indic calendar dates were used in mathematical texts in Indic languages, while Arabic, Persian, or Urdu texts could use regnal years of individual rulers in addition to the lunar Islamic, calendar. The most complete colophons included geographical, institutional, and social information about locations, schools, libraries, or patrons.

All of these types of statements are not found in manuscripts from the start of any of the mathematical traditions covered in this chapter; they emerged at different times and in different parts of the mathematical disciplines. In geometry, numbered or verbal references to previously formalized statements as listed above dominate in Arabic, Latin, and Hebrew translations of Euclid's *Elements*. They also can be found in the respective translations of other geometrical and astronomical works from antiquity, among them the works of Archimedes, Apollonius, Ptolemy, Menelaus, Autolycus, Theodosius, and Hypsicles. The earliest extant Greek copies of those texts, dating mostly to the ninth century, do not always possess all those features, while Arabic copies include them as early as the middle of the tenth century. In newly composed mathematical texts similar forms of structural organization can be found; the names of translators or editors now replaced by those of authors or commentators.

Mathematical texts composed in Latin, or vernaculars derived from it, shared with other works in those languages the development of their structural elements. Many were written in cursive hands in locally available script forms, but those produced for members of the nobility, or at particular monasteries, were also artfully designed. These "luxury" works were rich in colors and beautiful calligraphy—above all the titles, initials, or headlines—and contained carefully drawn diagrams or, in the later Middle Ages, even beautiful, mostly small-scale, paintings. For example, one copy of Gerard of Cremona's Latin translation of Menelaus' *Spherics* has long, sweeping initials and no diagrams, and breaks off after only a few propositions, so that it seems to have been made more for its cultural value than for its mathematical content (London, British Library, Harley 13, fol. 195r).

These sorts of embellishments were applied in particular to astrological texts and calendars such as hour books, but—as said above—can also be encountered in connection with practical mathematical problems. Tables of contents, foliations, and enumerations of quires were often added later. In cases of long or luxury manuscripts, they could be included by the scribe of the text or collection of texts. Roman numbers, Latin capital letters, and, in the later Middle Ages, also decimal place-value numbers were used in such cases. But there are also mathematical manuscripts extant that contain tables of contents or foliation as original structural features.

A specific kind of visual representation were initial letters beginning a text or a paragraph. Such letters were usually much larger than the ordinary size of the script and were often richly colored or even embellished with images of divine or human figures, including themes related to mathematics, astronomy or astrology, floral ornaments, or animals. A Latin multiple-treatise manuscript with texts for the teaching of the disciplines of the seven liberal arts in the trivium and quadrivium as taught at the University of Paris in the early fourteenth century contains many such initials of high artistic quality. Among them are representations of geometry drawn into the first letter of Book I of Euclid's *Elements* in the translation from Arabic by Adelard of Bath (London, Burney 275, fol. 293r), arithmetic and music at the beginning of Boethius' *On Arithmetic* and *On Music* (ibid., fols. 336r, 359v), respectively, and astronomy opening the introduction to Gerard of Cremona's translation of Ptolemy's *Almagest*, which was derived from two Arabic translations from the ninth century (ibid., fol. 390v). Ptolemy and Euclid were occasionally also portrayed in such initials. An example is the initial at the beginning of a copy of Ptolemy's *Almagest* that was made in 1213 from a manuscript in the library of the Abbey of St. Victor in Paris (Paris, BnF, lat. 16200, fol. 1r). Ptolemy is represented there as one of the kings of the Ptolemaic dynasty in Egypt.

Some Hebrew Bibles appropriated decorative astronomical motifs. Arabic, Persian or Ottoman Turkish mathematical, astronomical, or astrological texts were instead decorated with scripts in several colors and titles or headlines in particular, larger-sized calligraphic scripts.

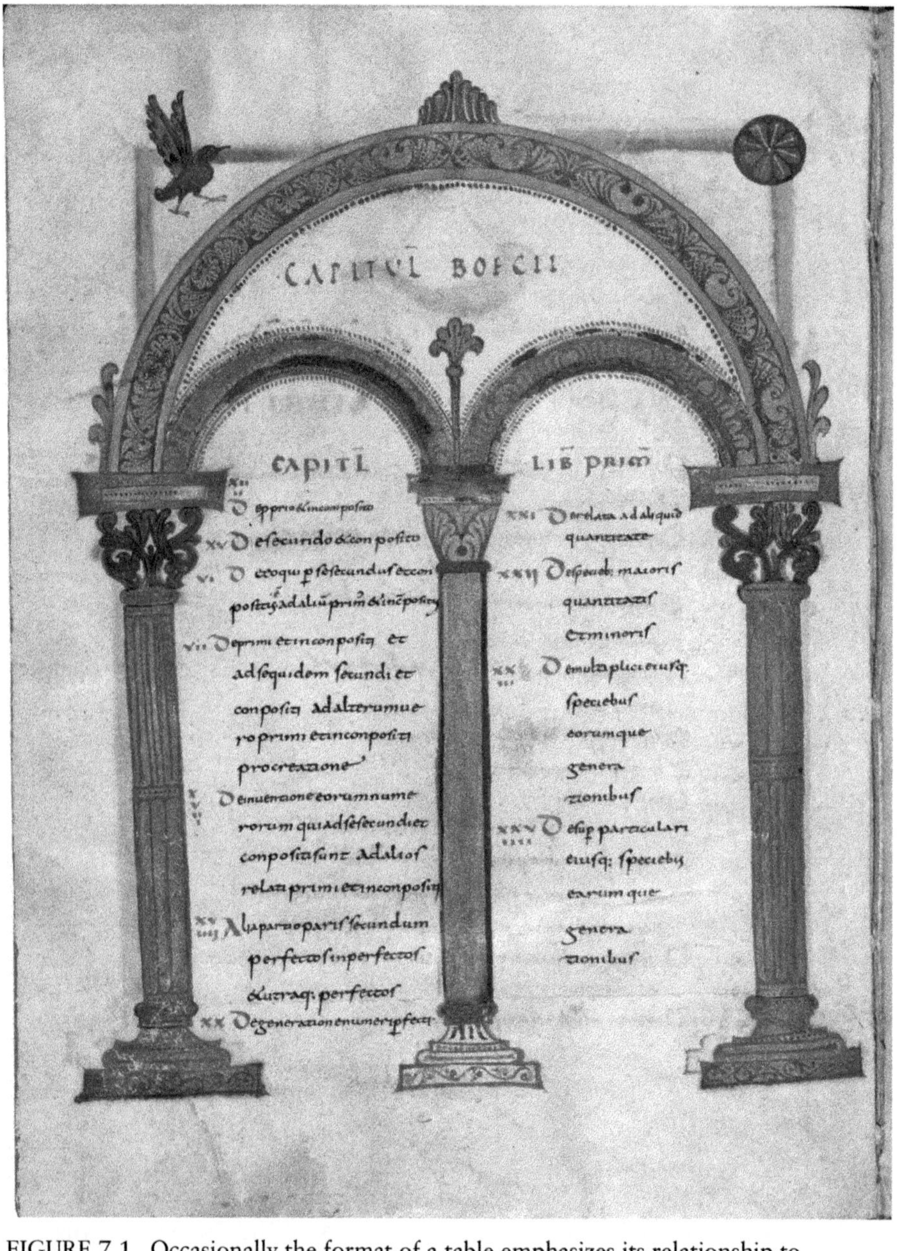

FIGURE 7.1. Occasionally the format of a table emphasizes its relationship to liturgical and computus (Easter calculation) texts. A copy of Boethius' *On Arithmetic* from about 845 documents this connection (Bamberg, Staatsbibliothek, Msc. Class. 5, fol. 7r). Photo by Gerald Raab, courtesy of the Bamberg State Library, Bamberg, Germany.

DIAGRAMS AND ILLUSTRATIONS IN GEOMETRIC AND ASTRAL SCIENCES

The diagrams of plane and solid geometrical figures were often indicated by a formalized designation such as *figure* or *image* and could also include a number. In copies of the *Elements* across the various languages into which it was translated over the centuries such drawings could appear within the text itself or in the margins. When they appear in the text, they usually come at the end of the proposition itself, so that we often see a diagram on the same page as the enunciation of the following proposition. Diagrams that are within the text are drawn either (a) at roughly the same time as the text is copied, so that the diagram and text jostle up against each other, (b) before the text is copied, with the text wrapping tightly around the diagrams, or, most often, (c) after the text is written and set into place holders that are sometimes delineated by still visible lines or indentations. These placeholders have been left blank in some manuscripts—blank boxes where the diagrams should have been drawn. In rare cases, copyists collected all the diagrams in one copy together with the enunciations of the propositions or with their respective numbers only, leaving out the remaining textual explanations; sometimes they added diagrams of special cases in the margins without providing their textual correspondences. In some cases, diagrams from a different version of the same text are drawn either into the margins, or on a separate page (De Young 2005; Menelaus' *Spherics*, London, British Library, or. 13127, fols. 52a–56a). After the fifteenth century, the diagrams of each of the books of the *Elements* began to be assembled at the end of each book. In Western Europe, this separation appears to have resulted from the production constraints of printing. In South Asia and much later in West Asia, manuscript copyists picked up such format changes from imported Latin, English, or French books.

Geometrical diagrams in mathematical and astronomical texts exhibit a number of characteristics that were probably derived from ancient models, in particular "overspecification" and "indifference to metrical accuracy" (Saito and Sidoli 2012: 140–8). The term "overspecification" refers to the fact that the diagrams show more symmetry and regularity than is required by the text, while "indifference to metrical accuracy" simply means that the diagrams exhibit little regard for metrical accuracy. For example, general triangles are depicted by isosceles or equilateral triangles, parallelograms are depicted by rectangles, conic sections are depicted by circles, unequal lines are depicted as equal and the converse, skew lines are drawn parallel to the margins, and so on. In many cases there is no mathematical problem with these diagrams, they simply differ from modern tastes. There are cases, however, where the diagrams have almost certainly been copied incorrectly and contain mathematical errors, often produced by excess symmetry. In fact, it has been shown that there is a natural tendency to introduce symmetry when copying a diagram (Acerbi 2017: 244–6; Carman 2018, 2020), so that some of this regularity may simply be the result of the process of transmission. Nevertheless, such errors cannot explain all of the differences between medieval and modern diagrams, and it is also likely that there were different ideas about what kinds of information should be depicted visually, and how. In diagrams drawn by the hands of competent mathematicians or their students, although we rarely find mathematical errors, we still find differences from modern tastes in the depiction of regularity and generality.

It may be helpful to look at an example of a mathematically incorrect diagram, in Figure 7.2 (left), to better understand how the practice of copying diagrams could lead to errors. As a clear example of a diagram that must have resulted from copying errors, we may take a diagram from Ptolemy's *Almagest* V 17 (Carman 2018: 221). In Vat. gr. 180, a manuscript of the tenth century, a diagram of the moon's orbit that is used to compute corrections to the motion of the lunar epicycle, the line BΔ is depicted as a horizontal line, whereas in the text angle BZA is said to be 60°, and EH, not EZ, is asserted to be perpendicular to BΔ. It is clear that the copyists have made the diagram so symmetrical that it no longer correctly depicts the objects involved in the discussion. In this case, the correct figure is not more complicated than that in Vat. gr. 180, it simply depicts BΔ as a skew line. On the other hand, the diagram in the oldest manuscript of the *Almagest*, copied in majuscule in the ninth century, is badly drawn but not yet mathematically impossible (Paris, BnF gr. 2389, fol. 155v). In fact, however, in a late thirteenth-century copy of al-Ṭūsī's revision of the *Almagest* made by one of Ṭūsī's students from the master's own copy, we see this diagram drawn correctly (Tehran Sipahsālār 4727: 486) (Figure 7.2 (right)), whereas many other diagrams in this manuscript exhibit overspecification (Figure 7.3). Indeed. Indeed, anyone who understands the text will be able to correct the diagram, and we find the diagram correctly drawn in all of the Arabic versions of the *Almagest* that we have seen. Indeed, it is impossible to imagine that the diagram that Ptolemy originally drew looked like the one in Vat. gr. 180.

On the other hand, there are also clear indications that there was a certain visual aesthetic that differed from modern tastes. We can see this clearly in a copy made by Ibn al-Haytham of the Arabic translation of the *Conics* of Apollonius, in which all of the conic sections are depicted with circular arcs. For example, in *Conics* I.27, a circular arc is used to depict a parabola. In this case, the circular arc is not a visual depiction of a parabola, but simply some curve that can represent the parabola and which has the same configuration as required by the argument (Figure 7.4).

Since such circular arcs are consistently employed for non-circular curves throughout the text, and since Ibn al-Haytham understood the *Conics* as well as anyone, it is not possible to argue that these are simply the result of scribal ignorance or carelessness. Rather, we are dealing with diagrams that function

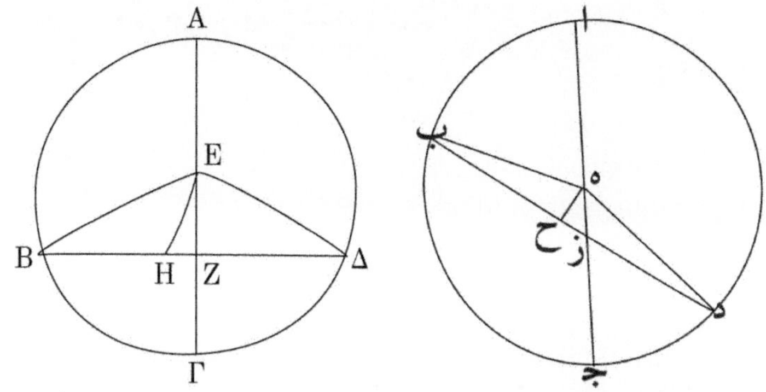

FIGURE 7.2 Reproductions of the third diagram of *Alm.* V.17, as found (*left*) in Vat. Gr. 180, fol. 129r (an incorrectly drawn diagram of the moon's orbit); and (*right*) in Tehran Sipahsālār 4727: 486 (the same correctly drawn). Reproductions by Nathan Sidoli.

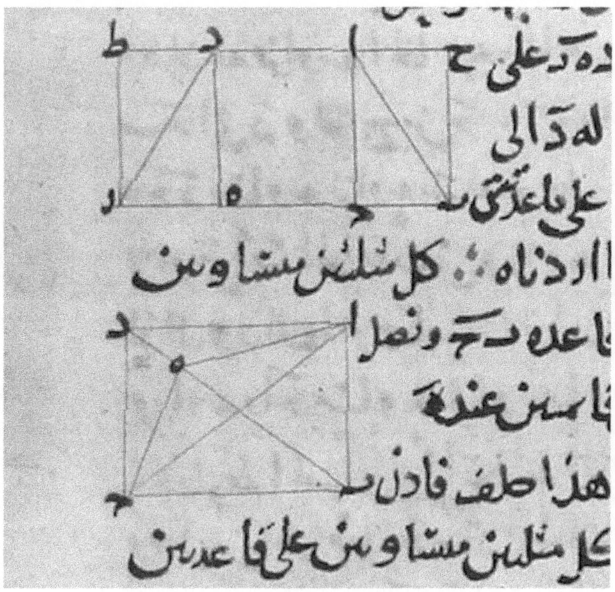

FIGURE 7.3. *Elem.* I.36, I.37 (Tehran Sipahsālār 4727: 14, copied from al-Ṭūsī's own copy). Courtesy of the Sipahsālār Library, Tehran, Iran.

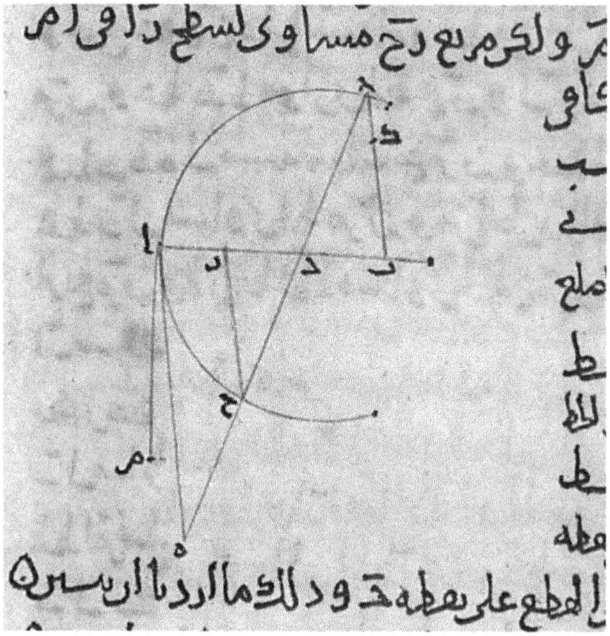

FIGURE 7.4. Parabola as a circle, *Conics* I.27 (Aya Sofya 2762, fol. 26a, 1024 CE, autograph of Ibn al-Haytham). Courtesy of Süleymaniye Manuscript Library, Istanbul, Turkey.

as schematic representations of mathematical information, not visual depictions of objects (Saito and Sidoli 2012: 157–60). That is, the medieval diagrams were meant to be used in conjunction with reading the texts as a way of facilitating some practice, such as teaching, learning, or doing mathematics.

Diagrams of solid configurations portray three-dimensionality in a rather schematic way. While some objects, such as the solids in the later books of the *Elements* are drawn in a way that is close to what we are familiar with seeing in modern diagrams produced using linear perspective, many solids are depicted more schematically. In particular, circles and ellipses that are viewed obliquely are depicted as lens-shaped figures (Figure 7.5).

As well as these diagrams, which give a visual sense of solid configurations, there were also techniques for flattening the image—some essentially visual, and others more mathematical—which were especially used for depicting spherical surfaces in the plane. In the transmitted and translated texts of ancient spherical geometry and astronomy, we find many diagrams in which the whole spherical surface has been opened up and flattened onto the plane, such that circles and arcs that are on the far side of the sphere are depicted outside of some base circle. This is, for instance, the case in the ninth-century Greek copy of Theodosius' *Spherics* (Vat. gr. 204, fol. 24r). In other texts, however, we find the elements of the sphere produced on the plane using drawing

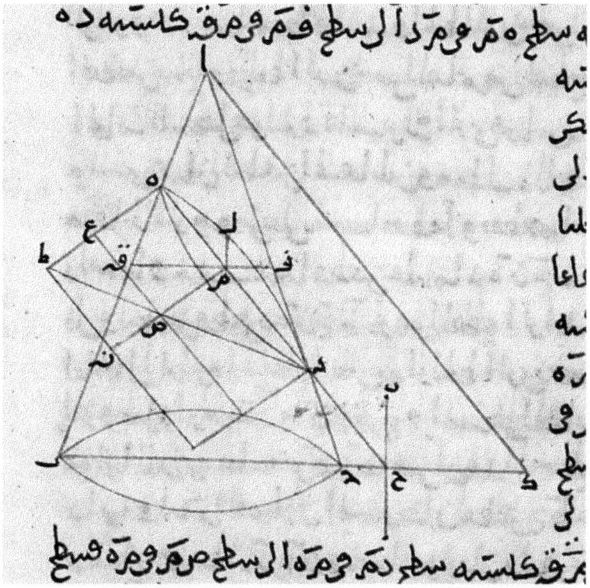

FIGURE 7.5. *Conics* I.13 (Aya Sofya 2762, fol. 16a (detail), 1024 CE, autograph of Ibn al-Haytham). Courtesy of the Suleymaniye Manuscript Library, Istanbul, Turkey.

techniques that are equivalent to various types of mathematical projection. For example, the so-called analemma diagrams use orthogonal projection into the receiving plane, such as in al-Bīrūnī's account of Ḥabash's method for determining the direction of Mecca (Leiden or., fols 168, 136b). Finally, many texts describe technical methods that can be used to project different systems of circles on the sphere onto a plane surface, which set out the theory of the construction of instruments like the astrolabe (al-Farghānī's *On the Astrolabe*, Berlin Staatsb., fols 58, 57b).

The diagrams that were produced for presentation copies were often beautifully drawn, such as those found in al-Ṭūsī's version of Euclid's *Elements* made in the fourteenth century for Fātiḥ Sulṭān Meḥmed (Istanbul, Feyzulla 1359). But sometimes more attention was paid to their artistic quality than their mathematical accuracy and mistakes were made, for example in the copy of a Latin paraphrase of the *Elements* made in the Carolingian context (Vatican, Pal. lat. 1564).

The formal features of diagrams visualizing textual knowledge were not limited to plane geometry, but also characterized the various treatises on spherical geometry, optics, and surveying. Colors were introduced to demarcate different geometrical figures in manuscripts produced for elite customers. In

Christian Europe, problems on surveying began to be illustrated by representations in three-dimensional forms at the latest by the fourteenth century. The development of techniques of the fine arts increasingly impacted visualizations of mathematical content. In Muslim states in South Asia, in particular in productions for the Mughal court, similar trends emerged in the sixteenth century, but focused there primarily on the astral sciences. Persian translations of French texts on the mixed mathematical sciences made in the eighteenth century in southern India reproduce competently the changes in technical drawings that had taken place up to that time in Europe. Medieval Hebrew geometrical, astronomical, or astrological texts shared their modes of visual representations of mathematical content as a rule with those of the societies in which they were produced. The same applies to Armenian, Syriac, and Georgian geometrical, astronomical, or astrological treatises.

Arabic, Persian, Ottoman Turkish, Hebrew, Latin, and Romance astronomical texts explaining the structure of the world often contain illustrations representing the movements of each of the planets according to a geocentric cosmological model, the phases of the moon, and lunar and solar eclipses. They often include a hemispherical map of the Old World or a schematic map of the seven climes. Beginning probably in the tenth century, a planar depiction of the known universe was added. Other astronomical texts, in particular so-called handbooks, often consisted primarily of short pieces of texts and many tables. In the seventeenth century, copies of Arabic or Persian texts on the astrolabe, planetary theory, and astrological themes produced in Iran or South Asia were occasionally illustrated with images from 'Abd al-Raḥmān al-Ṣūfī's *Book on the Star Constellations,* discussed below. Arabic or Turkish texts copied in Ottoman cities could reproduce images and painting styles from Italian manuscripts or Dutch books. An example among them is a copy of the Arabic translation of Book VIII of Pappus' (fourth century) *Collection*, most likely produced in Istanbul in the sixteenth or seventeenth century (Istanbul, Süleymaniye Library, Aya Sofya 3624, fols. 52a–102a).

One type of diagram accompanying texts of astronomy and cosmography deserves further mention, because it seems not to have had much precedent in ancient models—although it is adumbrated in descriptive works such as Geminos' *Introduction to Astronomy*. Starting from at least the ninth century, texts of cosmography include diagrams that have verbal labels that directly identify the various elements of the image, which we may call *verbalized diagrams*. This technique allows a greater density of information in the visual image itself, with the effect that we can often understand much of what these diagrams convey without reading the accompanying text, whereas in the diagrams of ancient astronomical texts it is usually necessary to read the text carefully in order to understand the diagram. As an example of this new visual presentation of the medieval period, we may take a diagram of the solar theory in an anonymous

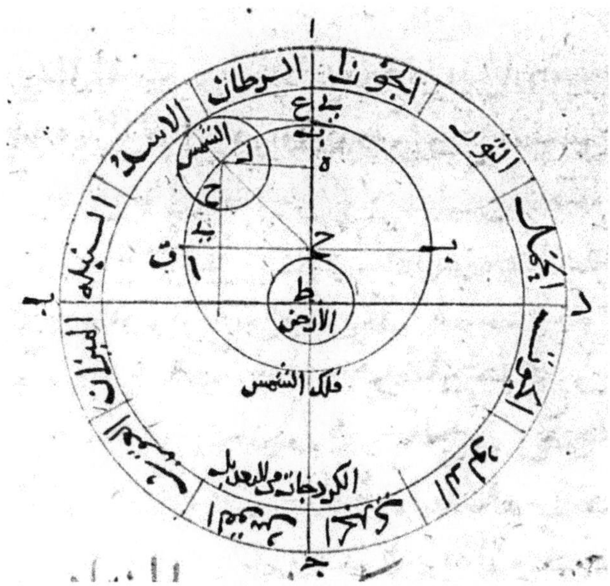

FIGURE 7.6. A diagram of the solar theory in an anonymous Arabic text of the late ninth to early tenth century with the names of the zodiac signs, the earth, the sun, and so on (Oxford, Bodelian Library, Arch Seld. A 11, fol. 56b). Courtesy of the Bodleian Library, University of Oxford, Oxford, UK.

Arabic text of the late ninth to early tenth century (Bodleian Arch. Seld. A, fol. 56b) (Figure 7.6). Written into the image itself we see labels for the earth, the sun, and the signs of the zodiac around the outer circle. In this way, we can see at once where the earth is, where the sun is, and where in the zodiac the sun's apogee lies, and so on. This style of diagram became a standard feature of the branch of astronomy known as *al-hay'a,* focusing on cosmography and orbital geometry, and later authors such as al-Ṭūsī used such diagrams to great effect (Tehran Sipahsālār 4727).

Another important class of diagrams that do not appear to have an ancient model are images that depict change over time. This is done in different ways in different disciplines and textual traditions. For example, in his *Memoir*, al-Ṭūsī provides a diagram to explain the complex motion of a pair of coupled circles set out in four panels, with descriptive text, having an equal time displacement between each panel (Tehran Sipahsālār 4727: 403) (Figure 7.7). This has the effect of reading across like the storyboard of a movie, or a comic strip, giving a visual impression of the passage of time.

A further class of such diagrams appears to have arisen in the medieval transmission of Pliny's *Natural History*, and then to have spread into related works on cosmology and time measurement. These diagrams show the latitudinal

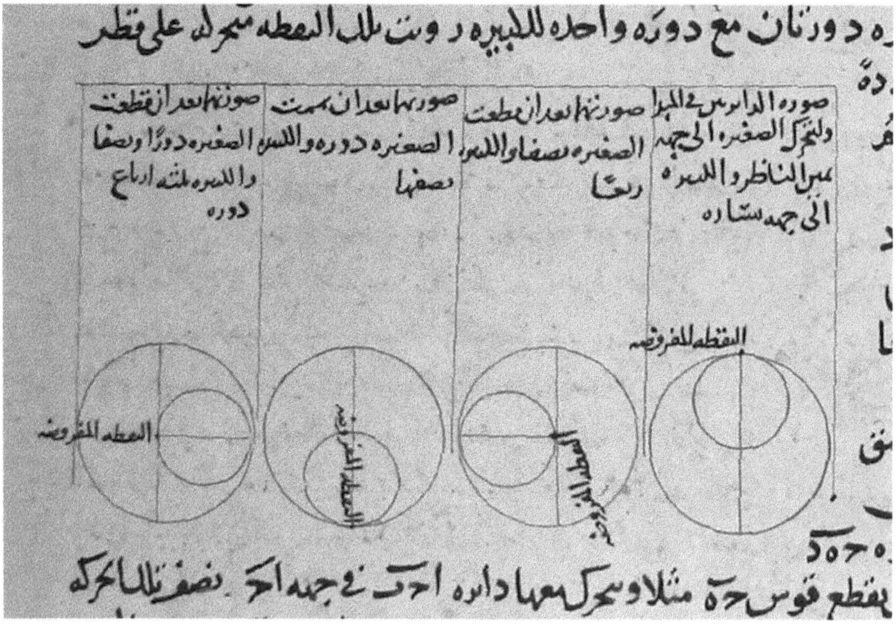

FIGURE 7.7. A four-part diagram demonstrating the motion of a pair of circles (Tehran Sipahsālār 4727: 403). Courtesy of the Sipahsālār Library, Tehran, Iran.

variation of the planets against their longitudinal position, which hence stands in for a variation in time. These diagrams are found in both circular and linear versions, and show the change in latitude as a sinusoidal curve, or as a linear zigzag function (Bern, Burgerbibliothek cod. 347, 25r; London, British Library, Roy 13.A.XI, fol. 143v).

Finally, in the beginning of the fourteenth century a new approach to the study of motion, and change generally, was developed by the so-called "Oxford Calculators" and disseminated throughout Western Europe over the following decades. Texts in a geometrical strain of this tradition, mostly practiced on the continent, involved diagrams in which various conceptual elements of change and motion are depicted. Among them are diagrams that we can recognize as proto-graphs, in which both of the axes and the area play some role in the visualization (ex. Munich BSB clm 18985, fol. 53r) (Figure 7.8).

Mathematical texts in Persian and various Indic languages can reflect the close cultural connections between different scholarly traditions under Muslim rule in South Asia. Although this is most often recognized for medical texts and astronomical instruments, mathematical texts also show some of the effects of the mutually shared scholarly and artisanal practices. Geometrical diagrams appear more often after 1600 in texts in Sanskrit and other Indic languages.

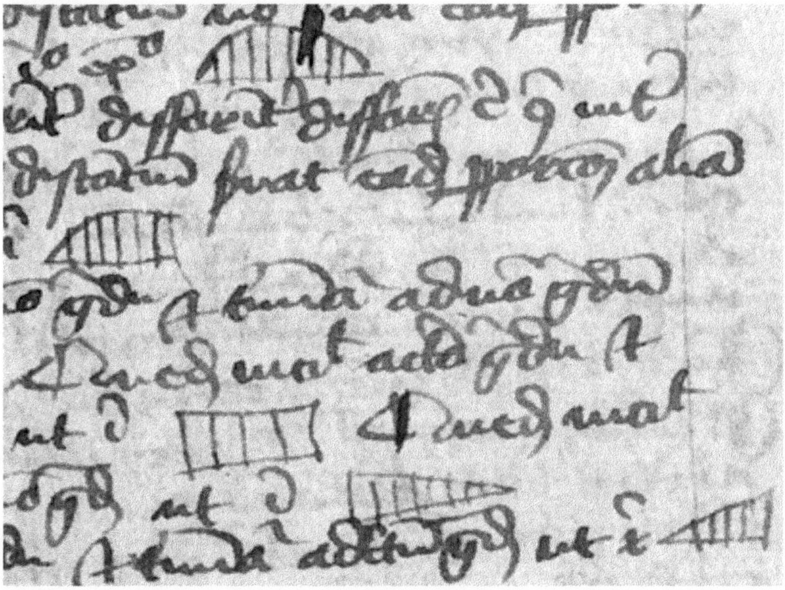

FIGURE 7.8. Diagrams of change and motion. (Munich, Clm 18985, fol. 53f). Courtesy of the Bayerische Staatsbibliothek, Munich, Germany.

NUMBERS, EQUATIONS, AND SPECIAL VISUAL METHODS

Translation was one of the core practices that promoted the spread and subsequent unification of both numerical signs and rules for the so-called fundamental arithmetical operations from South Asia to the west and the southeast. The translation of astronomical poetry as well as tables into Arabic during the eighth century opened the historical avenue for individual number symbols from zero to nine and the decimal positional system and its practices created in northern parts of the Indian subcontinent to become transformed into new variants as part of Arabic writing and calculation practices. Writing teaching texts and commentaries on these new signs and calculation practices was the second core cultural activity that contributed to their enduring and widespread success. Throughout the medieval period, these symbols were used alongside other ways of representing the numbers, such as Roman and *abjad* numeral systems, and locally traditional systems.

As outlined in Chapter 1, the numeral signs continued to be modified for several centuries after their arrival in Baghdad in the late eighth century. The slightly earlier encounter of the Syriac monk and author of astronomical treatises Severus Sebokht with the Indian numerical signs did not leave any substantial impact on the then widespread use of Greek, Syriac, Coptic, and possibly Hebrew letters for representing numbers in combination with shorter or longer strokes and dots. The Arabic copies of the Indian symbols, in contrast, achieved a profound impact in Islamicate, Christian, and Jewish societies that changed writing numbers and calculating in written forms up to the present.

These new signs were used in the eastern parts of the Islamicate world between what is today Pakistan and Egypt since the early ninth century, primarily in arithmetical texts, and in the later Middle Ages also in astronomical and astrological tables. Transported further to the west via North Africa, al-Andalus, Sicily, and perhaps other Mediterranean islands under Muslim rule, these originally Indian numerical signs underwent further modifications in their appearance. Used on "dust boards" covered with fine sand or soil for noting intermediary calculations and their results, they were called "dust numbers" early on in North Africa. Some of those North African forms continued in use after the translation of arithmetical texts into Latin in the twelfth century and stand at the beginning of today's number forms in scripts using the Latin alphabet. Other forms, like their predecessors, underwent multiple transformations before crystallizing into what we know today. Although dates were expressed verbally for a long time, in about the twelfth-century Indo-Arabic numbers began to replace the words for the years. Later, they were also used for the days.

Another important way to write numbers and calculate with them consisted in the combination of the Babylonian sexagesimal system with more locally traditional numeral systems, as mentioned above. Acquired through translations of Greek astronomical texts and tables into Latin, Syriac, and Arabic from Hellenistic times till the ninth century, and subsequently from Arabic into Persian, Hebrew, Latin, Armenian, Georgian, or Byzantine Greek and from Latin into other European languages starting in the tenth century, these forms of numerical representations were mainly employed in astronomical and astrological contexts.

Two other kinds of numerical representations were the Coptic and Roman numbers. Coptic number signs were used until about the thirteenth century in North African or Andalusian administrative documents. In Western and Central European Christian communities, Roman numbers were employed much longer and for a broader range of items such as calendars, the enumeration of quires, the calculi of the abacus, the calculation board, merchant diaries, trading contracts, or bank drafts. Multiplication tables taught with Roman numerals can be found, for instance, in copies of Boethius' *On Arithmetic*. An example is a manuscript from the third quarter of the twelfth century (MS London, British Library, Harley 549, fol. 14r).

In South and Southeast Asia, the forms of the numbers differed over time and among their material carriers. They changed with the development of new alphabets across the subcontinent. But the system and its rules as used in arithmetical or algebraic texts was primarily the decimal positional one. In tabular formats, they seem to be documented for texts composed from about the fifth century CE onwards. The histories of these multiple forms and their meanings and contexts have been investigated in very asymmetric modes that privileged Sanskrit and northern South Asia over the Dravida languages as well as languages derived from or mixing the one with the other (Keller 2006).

In traditions of number theory and arithmetic that were influenced by the Euclidean tradition, we often find numbers represented with diagrams consisting of a set of lines—either a single line per number, or two or more numbers laid out on a single line (Saito 2018). Although it is not strictly necessary to use these diagrams to follow the argument in the text, the diagrams are often arranged so as to facilitate understanding the argument and the mathematical objects involved, such as in Samaw'al al-Maghribī's treatment of what we would call the binomial expansion (Bajri, Hannah, and Montelle 2015: 559–68). Hence, it is clear that the use of lines to represent numbers was not simply a trope that carried over from geometrical texts, as is sometimes claimed, but was a way to organize and communicate various relations that exist between the numbers under discussion.

Special forms of numbers were the so-called figurate numbers which were represented through geometrical figures filled with an increasing number of dots starting from one and corresponding to the respective value of each number or, in the case of some Latin texts, with the Roman numeral "one" (MSS Madrid, BNE, 9088; London, British Library, Harley 549, fols 27r–v). An example are square numbers which were represented by an increasingly larger square to accommodate the series 1, 4, 9, 16, 25 . . . The origin of this kind of visualization comes from Greek copies of Nicomachus of Gerasa's (second century) *Introduction to Arithmetic* and such visualizations are also found in its translations into Latin, Syriac, and Arabic. Nicomachus' book also conveyed other kinds of visualizations, among them ratios, which were represented by arcs connecting the numbers in ratio.

Indeed, the details of computations with ratios are often worked out using visual arrays drawn into the margins of a number of other works, or explained using diagrams in the texts themselves. For example, we find the computation for removing one ratio from another worked out using Arabic numerals in the margin of a ninth-century copy of Ptolemy's *Almagest*, while the text simply states the result using the sexagesimal numerical system in Greek format (Vat. gr. 1594, fol. 26r). In al-Bīrūnī's text on the Indian Rule of Three he gives computational rules for working with compound ratios using a diagram to exhibit the various relations that can and cannot hold (Patna, Khuda Bakhsh Oriental Public Library 2591 (formerly Bankipore 2468), fol. 241a).

Arithmetical operations such as additions, multiplications, divisions, or extractions of roots were visualized in forms of quadratic or triangular tables, framed lists, lattices, or cross-like arrangements of lines connecting different operational elements. One very widespread form of multiplying two multi-digit numbers was executed in the form of a crosshatch or lattice configuration. It was known in South Asia, the Islamicate world, Western Christian Europe, and China, and carried thus a multitude of names, among them *shabaka* (net, grid), method of the sieve, *gelosia* (grille in front of a window) method, Venetian square, Chinese lattice, and other labels—now commonly called lattice multiplication (Figure 7.9). Depending on the number of digits in the two factors, the cells of a square or a rectangle are divided by diagonals connecting the upper right corner of each cell with the lower left corner. The same procedure was also arranged in the so-called diamond format, where the square or rectangle was rotated by 45°, or in a triangular format.

The method of double false position, also called the method of two errors, appeared early on in Chinese mathematical texts, for instance in Chapter 7 of the anonymous *Nine Chapters on the Mathematical Art*, entitled "Too Much and Not Enough," often translated as "Excess and Deficiency" (the two assumed

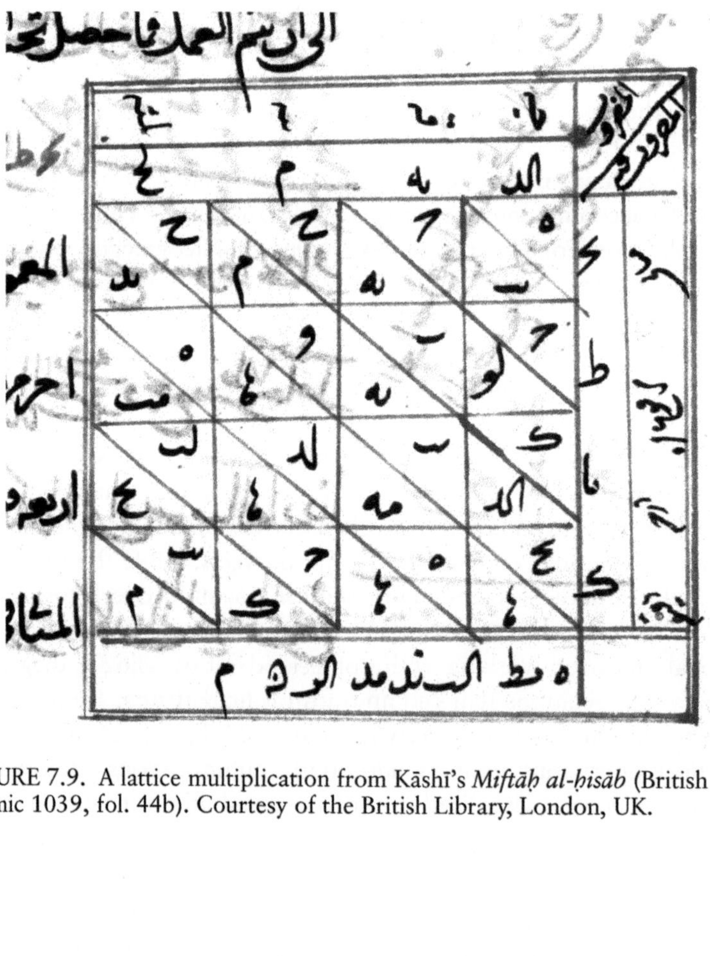

FIGURE 7.9. A lattice multiplication from Kāshī's *Miftāḥ al-ḥisāb* (British Library IO Islamic 1039, fol. 44b). Courtesy of the British Library, London, UK.

false values). A millennium and more later, it was often taught in Arabic and Persian school texts on arithmetic. This method for solving equations, attributed by Ibn al-Bannā' to Qusṭā ibn Lūqā, is called the method of the scales. Visualization can be found in many copies (see, for instance, Washington D.C., Library of Congress, H.1153, fol. 74b).

Diagrams for solving simultaneous polynomial equations in as many as four unknowns are found repeatedly in Chinese works. An example is an image from Zhu Shijie's 1303 *Jade Mirror of Four Unknowns* (Figure 7.10). It depicts what Needham refers to as "matrices" showing the ***tian yuan*** algebraic notation and the positions counting rods would assume from step to step as the algorithmic procedures to solve the problem were carried out. Needham explains that the "matrix" in the middle of the far right column in Figure 7.10 represents the equation $xy^2-120y-2xy+2x^2+2x$.

CELESTIAL GROUPINGS: CONSTELLATIONS, ZODIAC, PLANETS

The figurative representation of constellations, the zodiac, and the planets started in antiquity on clay tablets, seals, stone pillars, oil lamps, and globes as well as in manuscripts, mosaics, gemstones, tombs, temples, paintings on ceilings and elsewhere, and other material objects covering Mesopotamia, the Fertile Crescent, Egypt, Roman North Africa, Hispania, Italia, Gallia, Germania, and Britannia. Whether such representations found a home in ancient Iran is disputed. But it is clear that Greco-Roman visual representations of such celestial objects and astronomical concepts such as the Mesopotamian zodiac and the Egyptian decans reached northern South Asia in the late first millennium BCE and the first half of the first millennium CE. From there they spread further to the north, south, and east to be found as far east as Japan at the latest by the ninth century. The main conduit for Mesopotamian and Greco-Roman images of constellations, zodiacal signs, and planets from South to East Asia was the northward and eastward spread of Buddhism. Such images became staples of celestial representations in different Buddhist schools all over East and Southeast Asia as well as in Daoism in China. They underwent a number of changes reflecting developments within Indian contexts. Gemini, for instance, was transformed into a female-male pair and Libra was represented in several Buddhist and Daoist paintings as a Roman steelyard instead of the equal-armed balance. Particular components of Indian planetary representations such as the demon Rahu, responsible for eclipses and transformed into the ascending lunar node, or Ketu, originally a comet but later connected to Rahu and transformed into the descending lunar node, as well as the animals and other symbols ascribed to each planet and the sun, also travelled east and south in Asia.

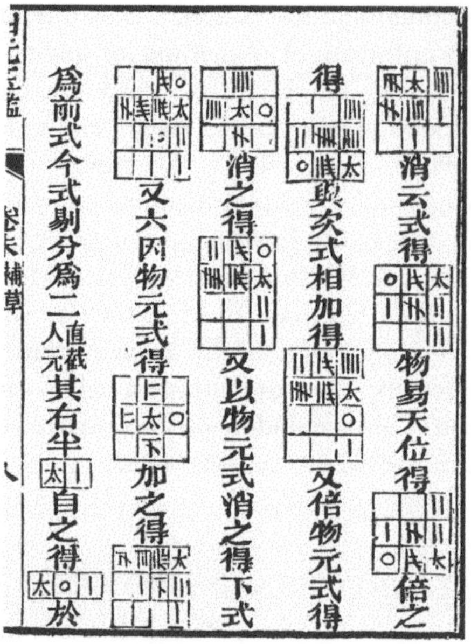

FIGURE 7.10. Counting rods representing coefficients of equations displayed on a "counting board" surface, from Zhu Shijie, *Jade Mirror of Four Unknowns* (1303) (Needham 1959: 131). Courtesy of the Needham Research Institute, Cambridge, UK.

Transmissions of such celestial representations westwards and northwards during the Middle Ages produced new forms of planetary images in Islamicate societies such as Venus with a harp or a flute, shown primarily in manuscripts produced for courts of Turkic dynasties, and a seven-armed Saturn with symbols combining Roman with seemingly Indian aspects but differing from the known Indian forms of Saturn. Modifications of the zodiacal signs occurred in Islamicate as well as Christian societies. Reflecting in all likelihood Chinese cosmological imagery, Gemini took the form of an intertwined pair sharing a snake body in a manuscript on poetry produced in Shiraz in the fourteenth century at the court of a smaller Mongol dynasty. More widespread became the representation of Gemini as conjoined twins in astrological manuscripts possibly produced in Baghdad at the court of another smaller Mongol dynasty (Oxford, Bodleian Library, March 144) and afterwards in Istanbul at the Ottoman court (Paris, BnF, Supplément turc 242, and New York, Pierpont Morgan Library, M.788). In Arabic, Persian, Turkish, Georgian, and Armenian astrological manuscripts Sagittarius might be represented from at least the twelfth century on with a feline body instead of the Mesopotamian horseman found in its ancient and medieval astronomical representations. An allegedly Indian component of this kind of Sagittarius that is believed to have passed through Late Antique Iranian transformations is the long tail ending in an open-mouthed dragon head at which Sagittarius shoots. This dragon head is thought to represent Rahu, the eclipse demon. The same imagery of Sagittarius appeared on coins minted during the twelfth and thirteenth centuries by Turkic dynasties in northern Iraq and Eastern Anatolia and on coins and medals struck in the early seventeenth century at the Mughal court in India. It can also be found on metal dishes, pen boxes, and inkwells produced from the twelfth to the fourteenth century in eastern Iran, northern Iraq, Syria, and Egypt, on bridges, fortresses, and other buildings in Syria, Anatolia, and Armenia from approximately the same time. Rahu and Ketu are believed to appear in astrological representations of Gemini on some of those metal objects. Their combination as the head and tail, respectively, of a snake or a dragon was a part of the iconography of the auspicious and inauspicious forms of astrological houses in manuscripts produced in the late fourteenth or early fifteenth century in Baghdad and in the late sixteenth century in Istanbul.

An important astronomical text exclusively dedicated to the discussion, numerical data, and figurative representations of forty-eight constellations presented by Ptolemy in the star catalogue of his *Almagest*, enlarged by a small number of star constellations ascribed to the Bedouins of the Arabian Peninsula, was ʿAbd al-Raḥmān al-Ṣūfī's *Book of the Constellations*. He compiled it in Shiraz in about 964 for his patron, the Buyid emir ʿAḍud al-Dawla, after having made observations to check the brightness and coordinates of a number of stars, critically analyzed contemporary writings about star positions and lunar mansions, and studied globes and treatises with figurative representations of the constellations. Some colophons in extant manuscripts report that he drew the figures himself in two copies of his work (Savage-Smith 2013: 137). Unfortunately, none of those copies survived. The earliest extant copies were produced in the twelfth century. They show a combination of Greco-Roman with Sasanian, ʿAbbāsid, and Turkic elements, in particular in the facial features, hairstyles, dresses, hats, jewelry, and other kinds of accessories.

Al-Ṣūfī's *Book of the Constellations* became the most often copied and illustrated mathematical book in the Islamicate world. The most recent copies were made in the later nineteenth century. Numerous constellations appear in multiple types, hinting at the repeated availability of additional Greco-Roman material. The extant Arabic and Persian copies range from high court art through provincial art to amateur copying of different skill levels. According to such variants, six different cycles of constellations have been so far identified in Arabic manuscripts (Duits 2005: 166).

Equally manifold as in Islamicate societies were the images of planets, zodiacal signs, other constellations, and the decans in parts of mediaeval Christian Europe. Ancient Mesopotamian, Egyptian, Greek, and Roman types were transmitted in copies of Latin translations of Aratus of Soli's hexameter *Phainomena* (3rd century BCE) and celestial maps included in some of them.

Such copies are extant in greater numbers from the eighth century onwards. An example is found in a collection of astronomical, astrological, and philosophical texts produced in 818 in the region around Salzburg (Munich, BSB, Clm 210).

The *Book of the Constellations* was also an important component of celestial representations in manuscripts produced in western Christian Europe. It is believed that a Latin translation was produced in Sicily, perhaps at the court of Frederic II. The surviving Latin exemplars with texts, tables, and images have been divided into eight different cycles. In other Latin copies, only the captions and figures are provided (Duits 2005: 166–7). Illustrations astoundingly close to those in Seljuq-style were for instance executed by an artisan working between 1250 and 1275, possibly in Bologna (Paris, BnF, Arsenal, 1036). Another translation was made into Castilian in the middle of the thirteenth century for Alfonso X of Castile and León.

A similar selective approach as in some Latin copies is documented in the Hebrew abridged version. It was part of an astronomical-astrological multiple-text manuscript made for Pedro IV of Aragon. Despite modifications and adaptations to local dress, numerous features of the iconography are undoubtedly appropriated in some more or less indirect way from some Arabic version of the work. Three incomplete and partially disordered copies are today known to exist (MSS University of Pennsylvania, Philadelphia, ljs 57: 116–144; Zurich, Braginsky Collection B93; Vienna, ÖNB, Heb. 132).

Images following representations known from Arabic copies of al-Ṣūfī's book also made it into independently composed astrological works and encyclopedias. Examples are Michael Scotus' thirteenth-century *Introductory Book* on astrology written for Frederic II and the encyclopedia of Prosdocimo de' Beldomandi from Padua (fifteenth century; Duits 2005: 171–3). Something similar happened in the sixteenth century in the Ottoman world, when Renaissance paintings of the constellations were included into a Turkish universal history (Istanbul, Süleymaniye, Lalaismail 318).

Books of Hours not only contained zodiacal signs and images of the labors of the months, but, being calendars, they also counted the days and months in words, Roman or decimal place-value numbers, and sometimes showed the course of the sun through the zodiac and the positions of the moon during the month. A splendid example for such illustrations is the hour book produced for the Duke of Berry.

Fascinating diagrams with calendrical and zodiacal content surrounding a figurative representation of a lunar bust can be found in mixed manuscripts as early as the eighth and ninth centuries. An example is a multiple-text codex from the St. Salvador and Bonifacius monastery in Fulda, now in the university library of Basel (Basel, University Library, F III 15a, fol. 22v) (Figure 7.11).

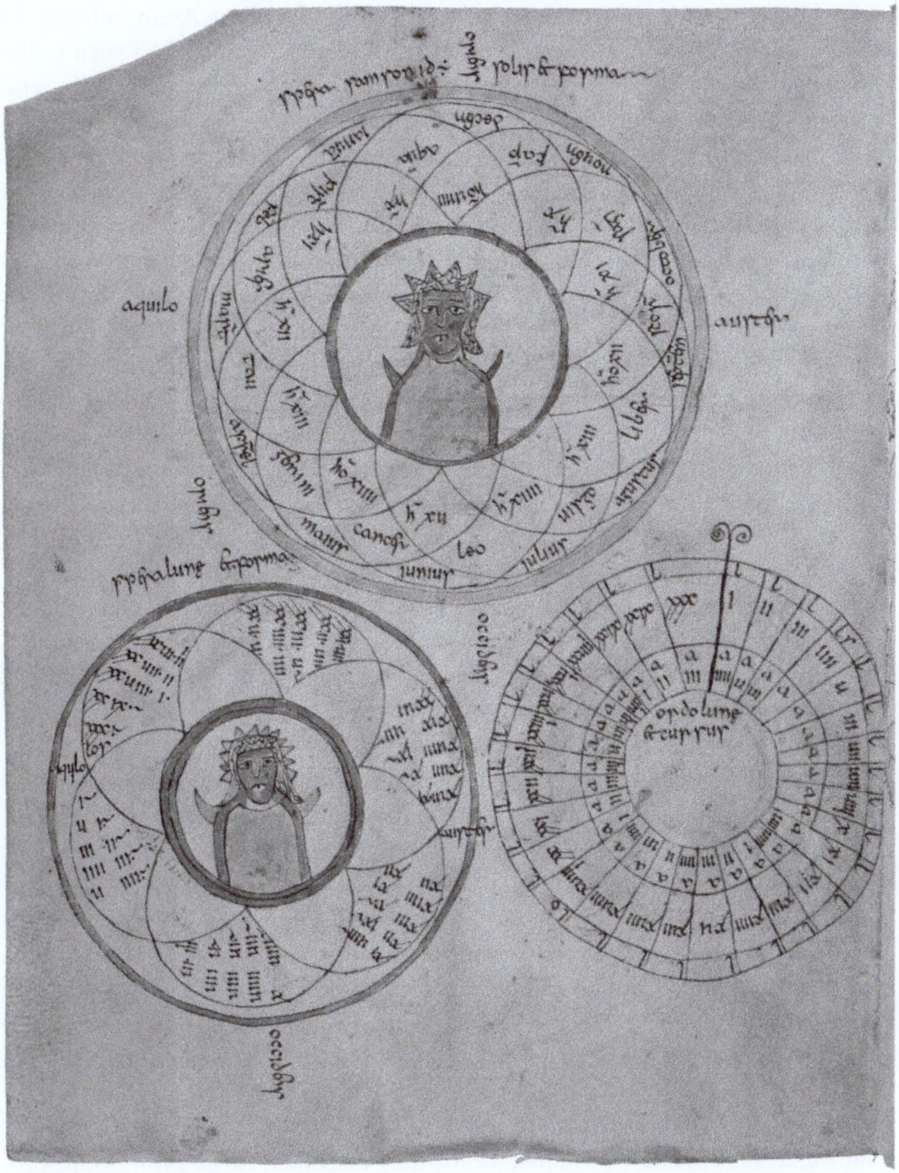

FIGURE 7.11. Diagrams with calendrical and zodiacal content surrounding a figurative representation of a lunar bust can be found in mixed manuscripts as early as the eighth and ninth centuries. An example is a multiple-text codex from the St. Salvador and Boniface monastery in Fulda, now in the Universitätsbibliothek Basel (UB Basel, F III 15a, fol. 22v). Courtesy of the University Library Basel, Basel, Switzerland.

Different, but equally variegated and rich are the East Asian cultures of celestial representations. Objects with individual star constellations such as the Big Dipper or the Pleiades or entire star maps from different pre-dynastic and dynastic periods from prehistoric times to modernity are preserved in territories today belonging to the People's Republic of China. They have been found in tombs, in Buddhist caves or temples, and on lacquer boxes and coffins, on ceremonial swords, and on stone stelae. The star maps could be circular or rectangular. The most famous rectangular star map was produced around 700 at Dunhuang (London, British Library, Or. 8210/S.3326). It contains about 1,300 stars painted in brown or black and red or orange. Several groups of asterisms were linked by straight lines forming a series of named constellations. Circular star charts could contain a few hundred additional stars. From the sixth century the first specimens are extant that include a representation of the Milky Way, called in Chinese the River of Heaven. During the reign of the Tang dynasty, circular star maps were occasionally included in gazetteers of locally collected information on a broad range of topics compiled for the imperial administration, but much more often this was the case under the Ming dynasty and its Manchu successor. Star maps following such Chinese ancestors can be found from the seventh century in Japan and the fourteenth century in Korea.

Many of those charts were placed on the ceilings of tombs, carved on stone stelae in front of temples, painted on silk scrolls hung in tombs, temples, or palaces, or painted on murals in the many Buddhist caves in northern China and northeastern Central Asia. In addition to the gazetteers, they are also found in manuscripts and printed books on divination in combination with the astral sciences. In these, depictions of comets, cloud formations, lunar mansions, and Buddhist or Daoist saints can also be found. Numerous objects and formats found in China and Japan, whether books, tomb ceilings, silk scrolls, maps, or mandalas, combine asterisms with lunar mansions, Indian-style zodiacal signs, and Buddhist saints from about the eighth century.

Lunar mansions, a partition of the orbit of the moon in twenty-seven or twenty-eight groups of asterisms and in exceptional cases a single star, are also known from South Asia. It is believed that the Indian twenty-seven, or twenty-eight, lunar mansions arrived in the Middle East at the latest in the eighth century, where, as twenty-eight stations, they became a further important tool for divination, in particular of weather phenomena. In Arabic manuscripts from Iraq, Anatolia, Syria, Egypt, North Africa, and al-Andalus, lunar mansions are usually represented by linear and curved strings of small circles filled with black or red ink and at times surrounded by a golden or differently colored circumference. In addition to manuscripts, they appear in circular diagrams, rectangular tables, or rings on Majorcan and Arabic portolan charts of the fourteenth, fifteenth, and sixteenth centuries. They are also depicted on some astrolabes as well as in paintings of astrolabes, for instance in an unfinished

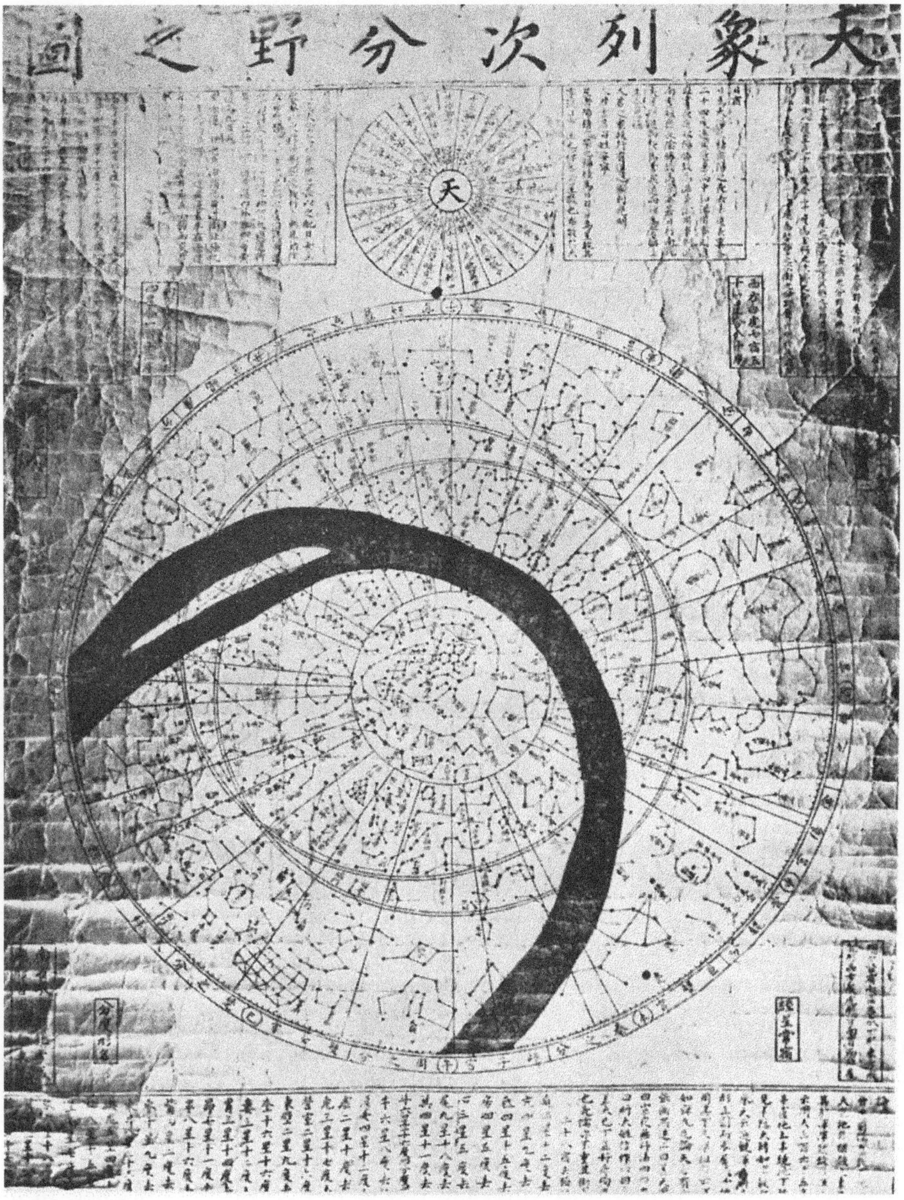

FIGURE 7.12. "Diagram of the Positions of the Heavenly Bodies and the Regions They Govern," a star map produced by a group of ten Korean experts in the astral sciences headed by Kwŏn Kŭn (1352–1409), (Needham 1959: plate XXVII, facing p. 280). Courtesy of the Needham Research Institute, Cambridge, UK.

painting of the back of an astrolabe component known as a *mater* in a Hebrew manuscript from the second half of the fourteenth century made for Pedro IV of Aragon (University of Pennsylvania, Philadelphia ljs 57: 111). Three astrolabes from Mosul and Damascus fabricated in the first half of the thirteenth century show them instead in figurative representations, as do three interconnected astrological manuscripts from late fourteenth- or early-fifteenth-century Baghdad and late sixteenth-century Istanbul mentioned above.

DIVINE AND HUMAN PRACTITIONERS OF MATHEMATICAL SCIENCES

While Eurasian mathematical traditions had much in common in their representations of mathematical content, including the astral sciences, in the art of portraying divine and human practitioners of those fields of knowledge more substantive differences occurred. The richest array of material in this regard comes from Western Christian Europe, followed closely by Chinese items. Islamicate and South Asian societies, in medieval times often closely interconnected, focused on a more limited set of themes.

Approximately from the thirteenth century onward, astrologers, followed later by geomancers, were depicted in miniatures painted in cities of Iraq, Iran, Syriac, Anatolia, Central Asia, and South Asia. These two kinds of practitioners of disciplines classified by numerous authors as belonging to mathematics provided the most common trope for representing a mathematical science. The images usually showed the practitioner holding some instrument and reading some information off it. The depictions were not portraits but standardized male figures in different postures clad in the characteristic robes of the region and time period. The most common setting of such paintings was the birth of a prince (as depicted in Figure 2.2). Such paintings in the Mughal tradition are abundantly extant for the sons of Mughal rulers from the sixteenth and seventeenth centuries and also for the births several centuries earlier of the Mongol ruler Ghazān Khān and Tīmūr, the founder of the Timurid dynasty (Worcester, Worcester Art Museum, 1935.12; London, V&A Museum, IS.2:80–1896; Dublin, Chester Beatty Library, In. 3, fol. 143r; London, British Library, Or. 12988, fol. 22r. et al.). But older paintings from the rule of the Mongol Ilkhanid dynasty in the thirteenth and fourteenth centuries also exist. In all those cases, the scenes decorate historical books. In the Mughal works, the painters often were Hindus, while in books from other Muslim dynasties they were mostly Muslims. Characteristic for most of these paintings is the cooperative nature of the enterprise: a group of scholars is reading texts, checking tables, reading off instruments, and talking to a ruler or a female administrator of the women's quarters. The purpose of those activities is the determination of the horoscope of the princely baby.

This function is also expressed by three astrologers together with Saturn holding two crowns in the sign of Gemini on the most beautifully illustrated extant horoscope of a Muslim prince—that of the Timurid ruler of Shiraz, Isfahan, and Yazd, Iskandar Sulṭān, cast on April 18, 1411 by the court astrologer Maḥmūd ibn Yaḥyā ibn al-Ḥasan al-Kāshī (London, Wellcome Institute, Persian 474).

Some images of the thirteenth and late fourteenth centuries, however, appear to portray individual scholars. Among them we find Aristotle holding an astrolabe and lecturing students, one of whom holds an open Greek book in uncial script (Istanbul, Topkapı Sarayı, Ahmet III, 3206, fol. 90a) and Abū Ma'shar holding an instrument, probably an astrolabe, in his right hand and putting his left hand on his head, apparently thinking deeply. While Aristotle is cast in the role of a teacher sitting before a group of students, Abū Ma'shar is the main actor in an unusual and uncertain pictorial story. Having shed his vestment except for trousers and a conical hat, he sits in a bowl on a pillar surrounded by a flower garden. A fully clad visitor—probably a scholar—looks at him with a finger of his left hand touching his lips, a standard Iranian gesture expressing astonishment. Parallel to Abū Ma'shar, we see his study with orderly stapled manuscripts, an inkwell, and his clothes. The headline of the image is heavily damaged at its beginning. Only the words "the philosopher Abū Ma'shar from Balkh" are clearly recognizable. A reader later added the incomplete information that Abū Ma'shar had been an astrologer in Baghdad in the time of Caliph al-Ma'mūn in the early ninth century (Oxford, Bodleian Library, Marsh 133, fol. 34a).

A rare collection of images of ancient philosophers is a manuscript held today by the Aga Khan Museum in Toronto. It was produced in Iraq at some time during the thirteenth century by the miniature painter Maḥmūd ibn Abī l-Maḥāsin al-Qashī in cooperation with the famous calligrapher Yaqūt al-Musta'ṣimī. Each folio shows a philosopher teaching a student. One of them is Pythagoras (Toronto, Aga Khan Museum, AKM 283, fol. 21b).

In South Asian non-Muslim contexts, astrologers are sometimes depicted in versions of the *Kalpasūtra*. These verses recount the life of the founder of the Jain community, Mahāvīra. In a nineteenth-century copy, Mahāvīra's father King Siddhārta as well as his first mother Devānandā are described as surrounded by *brāhmaṇas* responsible for rituals connected with the stars and dream diviners. Devānandā was Mahāvīra's first mother because he had descended from the heavens into her womb at a particular celestial moment, when the moon was in conjunction with the asterism Uttaraphalgunī. After some time, he left her womb under the rule of the same asterism and moved into the one of King Siddhārta's wife, Triśala, who gave birth to him when the time was right and Uttaraphalgunī reigned again. Mahāvīra entered the following cycles of his life under the ascendancy of the same asterism. Hence, it is not surprising that manuscripts of the *Kalpasūtra* showed various moments of Mahāvīra's life and that of his families as interpreted by practitioners of the astral sciences.

In the case of Siddhārta, an entire folio was dedicated to those priestly practitioners who worked at his court. Devānandā, however, is represented with her husband, Ṛṣabhadatta, who interpreted for her the fourteen symbolic objects she had seen in her dreams when Mahāvīra descended at night into her womb as an embryo, predicting a son with a myriad of miraculous attributes including the perfect knowledge of all the brahmanical sciences among which he also named arithmetic and the astral sciences.

In western Christian Europe, at least three types of figurative representations of the mathematical sciences evolved during the Middle Ages: Jesus Christ as connected with knowledge of the stars and geometry; symbolic or scholarly representatives of the four mathematical sciences of the quadrivium; and fictitious images of scholarly authors of the mathematical sciences from Greek and Roman antiquity, of authors of Arabic astrological treatises translated into Latin, and of patrons of the mathematical sciences. In the first class, Jesus Christ appears in different kinds of compositions as Pancreator. One such composition had its origins in representations of an ancient sun god in several parts of the Roman Empire, in which Helios, Sol invictus, or Mithra had been positioned in the center of various mosaics surrounded by the zodiacal circle. Even synagogues, such as the fourth-century Severus Synagogue in Hammath-Tiberias or the synagogue of Bet Alpha from the sixth century, show such a view of the cosmos (see Figure 2.5). The sun god rides in the center in a chariot drawn by four horses with the moon and stars behind him. This image is surrounded by the zodiac, where each sign is represented as an image and a name. In the four corners, images of the four seasons are placed (Englard 2003: 189–214; Fine 2005: 84–91, 196–205). Very similar kinds of composition with Helios replaced by Jesus Christ are found in early Byzantine churches and in later medieval monasteries on the Peloponnese and in the Iviron and Megistis Lavras monasteries on Mount Athos. They also can be found in medieval Latin manuscripts, sometimes still with a sun god or the sun itself in the center. There is a fifth-century Byzantine church showing the zodiac surrounding Helios in a mosaic on the floor in Maltezana on the Island of Astypalaia (Dodekanes).

A variation of this basic theme, which has often been considered a calendrical representation, appears in a drawing called the Calendar of Zwiefalten from about 1145 (Stuttgart, Würtemberische Landesbibliothek, Cod. hist. fol. 415, fol. 17v). In this drawing, the year as regent of time is placed in the center holding the sun and moon in his hands, beneath which day and night are represented. The year is surrounded by the zodiac followed by a ring with the labors of the months. Outside the ring of the months, twelve winds are added. In the four corners are images of the four seasons. In the corners of the upper and lower margins, four persons are placed who represent dawn, noon, evening, and then, surprisingly, the winter solstice.

In lavishly illustrated collections of moral-providing biblical stories, called *Bible moralisée*, the Pantocrator is depicted with a compass in one hand (as in Figure 4.6). The globe can be held in the other or placed on the ground. The globe in the hand is probably a surviving symbol of *Sol invictus*. Such codices were made for royalty from the thirteenth to the fifteenth century, starting with exemplars made for the French King Louis IX on order of his mother Blanche of Castile. Not all of the extant copies contain a miniature of the Pantocrator. Those that do vary in complexity and style, while always showing him as operating with a compass on the earth (examples of the thirteenth century: Toledo, Cathedral, The Bible of St. Louis, vol. 1, fol. 1v; Vienna, Österreichische Nationalbibliothek, Codex Vindobonensis 1179, fol. 1v; Vienna, Österreichische Nationalbibliothek, Codex Vindobonensis 2554, fol. 1v; Oxford, Bodleian Library, Bodley 270b, fol. 1v).

The symbolic representation of the disciplines of the quadrivium in medieval manuscripts, paintings, pillars, and other architectural contexts also continues Roman and Christian material and textual traditions. The literary context of such representations was Martianus Capella's fifth-century work *On the Marriage of Philology and Mercury*. Beautiful paintings of the four mathematical arts adorn many of the more than 200 medieval copies of the work in Latin and in translations. Examples are the series in a multiple-text manuscript today preserved in Munich from the thirteenth century and a Venetian manuscript produced between 1485 and 1490 (Munich, BSB, Clm 2599, fols. 101v–111r; Venice, Biblioteca Marciana, Lat. XIV, 35). The Munich manuscript contains a series of paintings in which the disciplines of the seven liberal arts are paired with ancient authors, as well as double author portraits. Boethius joined arithmetic and music, Cassiodorus geometry and Ptolemy astronomy. The Venetian manuscript, illuminated by Attavante degli Attavanti, was made in Florence for Matthias Corvinus, the King of Hungary and other regions in Central and Southern Europe. Boethius' *On Arithmetic* also propagated the concept of the quadrivium. Unsurprisingly, a number of copies of this work translated the idea into a miniature. An example is the manuscript previously shown in Figure 7.1, produced in Tours around 845 (Figure 7.13; Bamberg, Staatsbibliothek, Msc. Class. 5, fol. 9v).

Martianus Capella's allegorical *On the Marriage* became a rich source for depictions of the mathematical disciplines in the medieval Christian world (Stahl 1977). This work describes a divine wedding, in which the groom, Mercury, presents his bride, Philology, with seven sisters, who are the classical liberal arts. Each is introduced in turn and discourses on her subject. When Geometry appears, she is accompanied by a group of women who carry an abacus board, covered with green dust for drawing geometric diagrams.

FIGURE 7.13. The disciplines of the mathematical quadrivium as depicted in a copy of Boethius' *On Arithmetic*, painted in Tours around 845 (Bamberg, Staatsbibliothek, Msc. Class. 5, fol. 9v). Photo by Gerald Raab, courtesy of the State Library, Bamberg, Germany.

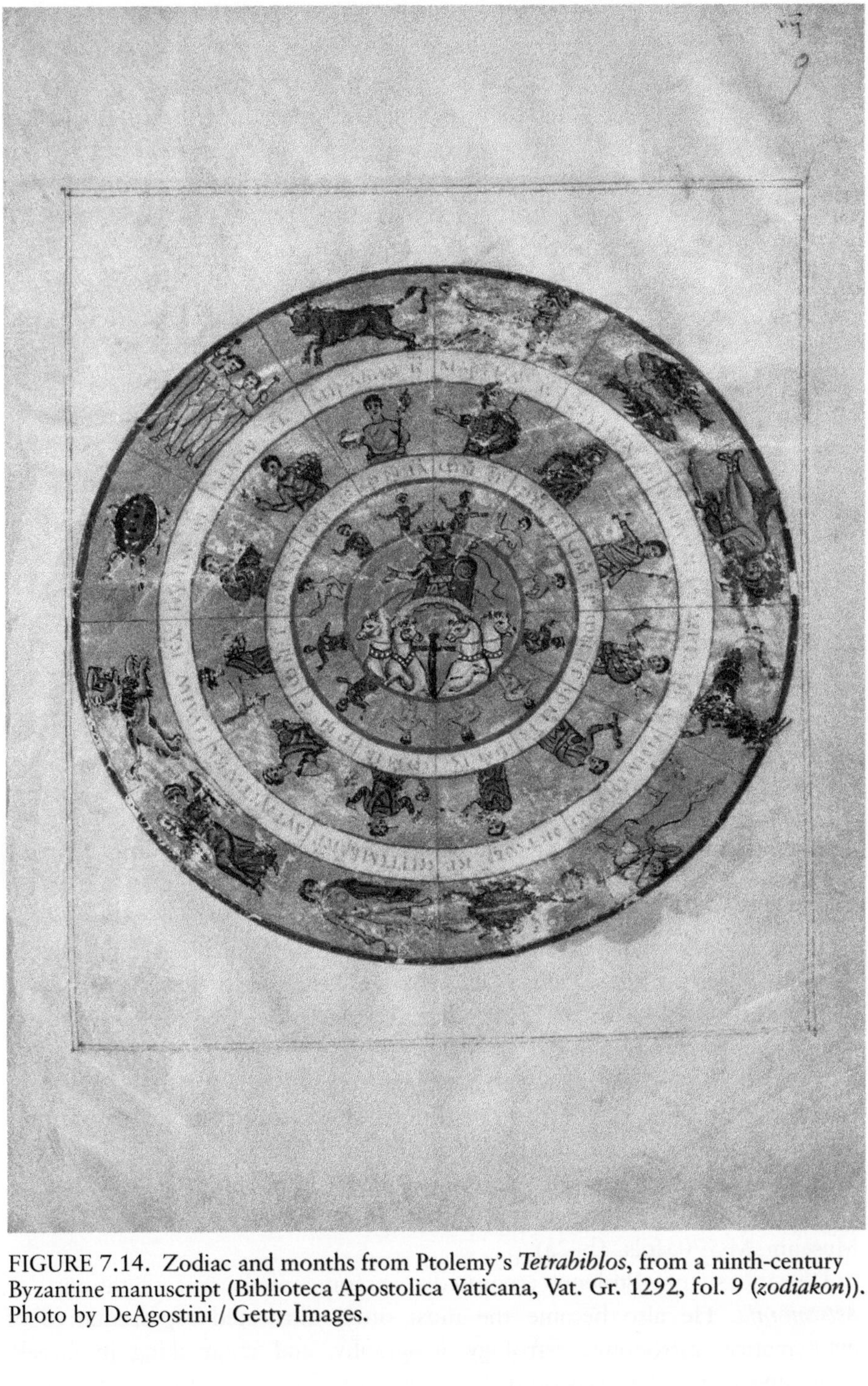

FIGURE 7.14. Zodiac and months from Ptolemy's *Tetrabiblos*, from a ninth-century Byzantine manuscript (Biblioteca Apostolica Vaticana, Vat. Gr. 1292, fol. 9 (*zodiakon*)). Photo by DeAgostini / Getty Images.

Geometry herself is described as a distinguished-looking lady, with a geometer's rod in one hand and an engraved globe, or celestial sphere, in the other. Her clothing is the color of the vernal sky, embroidered with numbers, gnomons, weights and measures, and figures depicting intervals. When Arithmetic appears next, the goddess Athena instructs that the abacus should be kept in place for writing numbers. Arithmetic herself is described as striking in appearance, and ten luminous rays emanate from her forehead. She performs finger calculations, and her dress is said to conceal the functions of nature and covers an intricate undergarment. Astronomy enters the scene in an illuminated sphere, surrounded by lesser deities, and impressing the crowd with her beauty. Her clothes and hair are covered with gems and she holds an observational instrument in one hand and a book containing astronomical tables in the other. The entrance music for Harmony, the last of the bridesmaids, is played by various gods and poets. Harmony herself is described as lofty, with gold adorning her head and wearing a garment stiff with embroidered gold. She holds in one hand a number of models of instruments and in the other a circular shield covered in circles, from which emanates music that silences all the rest.

Stelae of the mathematical arts are known from monasteries or clerical houses attached to cathedrals in twelfth-century southern France and perhaps also from elsewhere (Autun, Musée Rolin, inventory number 983.14.1 (geometry) and 983.14.2 (astronomy)). The four mathematical disciplines also appear on paintings, murals, tapestry, woodcuts, and prints, in particular from the fourteenth to the sixteenth centuries. *Arithmetica* is often depicted as a member in a party of merchants calculating with various instruments.

Individual human representatives of the mathematical sciences are known from Latin and later also vernacular manuscripts in Western Europe starting from around the ninth century. The scholars were represented as authors of well-known treatises, in particular Euclid's *Elements*, Ptolemy's *Almagest* or *Tetrabiblos*, or Boethius' *On Arithmetic*. The abovementioned Tours manuscript of Boethius' *On Arithmetic* shows Boethius and his father-in-law, the politician, philosopher, and historian Symmachus, as an author portrait before the frontispiece (Bamberg, Staatsbibliothek, Msc. 5, fol. 2v).

Medieval author portraits continue ancient representational traditions from antiquity, as similar author portraits (mostly medical) in Arabic manuscripts confirm. An exceptional double-sided author portrait of two astronomers thought to represent Ptolemy and ʿAbd al-Raḥmān al-Ṣūfī, or al-Ṣūfī and his son, opens an Arabic manuscript of the twelfth century (Tehran, Reza Abbasi Museum, M 570, fols. 1b–2a).

Depictions of Ptolemy repeatedly accompanied representations of *Astronomia*. He also became the most often illustrated representative of mathematical astronomy, astrology, geography, and mapmaking in Greek, Latin, and vernacular texts on different sciences, history, and religious works.

An example in a Greek manuscript of the fifteenth century is housed in the National Library of France. Together with a medieval scholar slowly moving a globe with a grid and showing the African continent according to the latest state of knowledge, Ptolemy in ancient dress holds a compass and a quadrant in his left hand while pointing with his right hand up to the heavens, where the Great Bear and Little Bear appear in the figurative imagery of the abovementioned works on the star constellations. The model of the two bears might even have been an Arabic copy of al-Ṣūfī's book, because they are portrayed with the typical long tails of the Oriental modes of painting (Paris, BnF, Grec 1410, fol. 2r).

Ptolemy also appears together with Abū Ma'shar, the eleventh-century Egyptian scientist 'Alī ibn Riḍwān, and King Alfonso X on the dials of astronomical clocks in late fourteenth- and mid-fifteenth-century churches on the Baltic Sea. Their images point to the shared visual and astral knowledge among the craftsmen involved.

Euclid is depicted repeatedly either in initials to the *Elements* or in miniatures in other scholarly works, among them writings about divination. Copies of a poem on fortune predictions attributed to the twelfth-century Bernardus de Silvestris, for instance, present Euclid together with Hermann (of Carinthia?) in colored drawings. Euclid holds a diopter to his eye watching the sky and an armillary sphere in his other hand, while an astrolabe hangs down from Hermann's right hand (Oxford, Bodleian Library, Ashmole 304, fol. 2v; Digby 46, fol. 8v).

Manuscripts of Germanicus' Latin translation of Aratus' *Phaenomena*, as well as transmitting images of globes and planispheres, contain an image depicting Aratus pointing at a globe with a rod or stylus, while Urania, the muse of astronomy, stands by with a bookroll or codex (Madrid Bib. Nac. 19, 55r; Aberstwyth NLW MS 735C, 11v). This image continues an ancient trope of a seated figure interacting with a celestial globe with a stylus, of which there are many examples from antiquity in various media—the figure may be male or female and may carry a roll or a codex, but the globe and the pointer, or stylus, are constant.

Pythagoras was another ancient Greek scholar who was repeatedly represented in medieval artwork and mathematical, musical, or educational manuscripts. He is honored with a sculpture over the western portal of the Cathedral of Chartres; he plays various musical instruments, in one case together with Philolaus, in a 1492 manuscript of the *Theory of Music* by Franchino Gafurio; and he turns away in utter disgust from a drawing of broad beans in a *Mirror for Princes* from 1512/4, rejecting the consumption of that vegetable (Washington, D.C., National Gallery of Art, 2006.11.4, fol. 25r).

BIBLIOGRAPHY

Acerbi, Fabio (2017), "The Mathematical *Scholia Vetera* to *Almagest* I.10–15 with a Critical Edition of the Diagrams and an Explanation of Their Symmetry Properties," *SCIAMVS*, 18: 133–259.

Al-Hassani, Salim (2016), "The Journey of Automatic Machines in Muslim Civilisation," *Muslim Heritage*, October 24. Available online: https://muslimheritage.com/journey-of-automatic-machines/#cas (accessed May 8, 2023).

Āpaṭe, V. G. (1937), *Līlāvatī*, Ānandāśrama Sanskrit Series, #107, 2 vols, Puṇe: Ānandāśrama Press.

Bagheri, Mohammad (1997), "A Newly Found Letter of al-Kāshī on Scientific Life in Samarkan," *Historia Mathematica*, 24 (3): 241–56.

Bajri, Sanna, John Hannah, and Clemency Montelle (2015), "Revisiting Al-Samaw'al's Table of Binomial Coefficients: Greek Inspiration, Diagrammatic Reasoning and Mathematical Induction," *Archive for History of Exact Sciences*, 69 (6): 537–76.

Barney, Stephen A., W. J. Lewis, J. A. Beach, and Oliver Berghof (2006), *The Etymologies of Isidore of Seville*, Cambridge: Cambridge University Press.

Baxter, William T. (1989), "Early Accounting: The Tally and Checkerboard," *Accounting Historians Journal*, 16 (2): 43–83.

Bell, John Lane (2006), *The Continuous and the Infinitesimal in Mathematics and Philosophy*, Milan: Polimetrica.

Berger, Anna M. B. (1990), "Musical Proportions and Arithmetic in the Late Middle Ages and Renaissance," *Musica Disciplina*, 44: 89–118.

Berggren, J. L. (1986), *Episodes in the Mathematics of Medieval Islam*, New York: Springer.

Berggren, J. Lennart (2007), "Mathematics in Medieval Islam," in Victor J. Katz, ed., *The Mathematics of Egypt, Mesopotamia, China, India, and Islam: A Sourcebook*, 515–675, Princeton, NJ: Princeton University Press.

Biggs, Norman (2009), "Mathematics of Currency and Exchange: Arithmetic at the End of the Thirteenth Century," *BSHM Bulletin: Journal of the British Society for the History of Mathematics*, 24 (2): 67–77.

Blair, Sheila, and Jonathan Bloom, eds. (2013), *God is Beautiful; He Loves Beauty: The Object in Islamic Art and Culture*, 122–55, New Haven, CT: Yale University Press.

Bork, Robert Odell (2016), *The Geometry of Creation: Architectural Drawing and the Dynamics of Gothic Design*, Abingdon: Routledge.

Boyer, Carl (1959), *The Rainbow: From Myth to Mathematics*, New York: Yosellof.

Branson, William B. (2013), "Solving the Cubic with Cardano," *Convergence*. Available online: https://www.maa.org/press/periodicals/convergence/solving-the-cubic-with-cardano-introduction (accessed July 2, 2023).

Brentjes, Sonja (2018), *Teaching and Learning the Sciences in Islamicate Societies (800–1700)*, Turnhout: Brepols.

Burnett, Charles (2001), "The Coherence of the Arabic-Latin Translation Program in Toledo in the Twelfth Century," *Science in Context*, 14 (1/2): 249–88.

Burnett, Charles (2006), "The Semantics of Indian Numerals in Arabic, Greek and Latin," *Journal of Indian Philosophy*, 34 (1/2): 15–30.

Burnett, Charles (2009), "Manuscripts of Latin Translations of Scientific Texts from Arabic," *Digital Proceedings of the Lawrence J. Schoenberg Symposium on Manuscript Studies in the Digital Age*, 1 (1): art. 1. Available online: https://repository.upenn.edu/ljsproceedings/vol1/iss1/1/ (accessed July 2, 2023).

Cajori, Florian (1906), *An Elementary History of Mathematics*, London: Macmillan.

Cajori, Florian (1928), *A History of Mathematical Notations*, vol. 1, La Salle, IL: Open Court Publishing Co.

Cantor, Moritz Benedikt (1900), *Vorlesungen über Geschichte der Mathematik, vol. 2: 1200–1668*, 2nd edn, Leipzig: Teubner.

Carman, Christián C. (2018), "Accounting for Overspecification and Indifference to Visual Accuracy in Manuscript Diagrams: A Tentative Explanation Based on Transmission," *Historia Mathematica*, 45 (3): 217–36.

Carman, Christián C. (2020), "Vestiges of the Emergence of Overspecification and Indifference to Visual Accuracy in the Mathematical Diagrams of Medieval Manuscripts," *Centaurus*, 62 (1): 141–57.

Chabás, José, and Bernard R. Goldstein (2017), *A Survey of European Astronomical Tables in the Late Middle Ages*, Leiden: Brill.

Chabert, Jean-Luc (1999), *A History of Algorithms: From the Pebble to the Microchip*, Berlin: Springer.

Chemla, Karine (2016), "Numerical Tables in Chinese Writings Devoted to Mathematics: From Early Imperial Manuscripts to Printed Song-Yuan Books," *East Asian Science, Technology, and Medicine*, 44 (1): 69–122.

Chemla, Karine, ed. (2012), *The History of Mathematical Proof in Ancient Traditions*, Cambridge: Cambridge University Press.

Chen, Qun Rene (2009), "Cardinal Directions in Chinese Language: Their Cultural, Social and Symbolic Meanings," *ETC: A Review of General Semantics*, 66 (2): 225–39.

Chrisomalis, Stephen (2010), *Numerical Notation: A Comparative History*, Cambridge: Cambridge University Press.

Clagett, Marshall (1953), "The Medieval Latin Translations from the Arabic of the *Elements* of Euclid, with Special Emphasis on the Versions of Adelard of Bath," *Isis*, 44 (1/2): 16–42.

Cullen, Christopher (2009), "People and Numbers in Early Imperial China," in E. Robson and J. Stedall, eds. *The Oxford Handbook of the History of Mathematics*, 591–618, Oxford: Oxford University Press.

Cullen, Christopher (2016), *The Foundations of Celestial Reckoning: Three Ancient Chinese Astronomical Systems*, London: Routledge.

Cullen, Christopher (2017), *Heavenly Numbers: Astronomy and Authority in Early Imperial China*, Oxford: Oxford University Press.

Datta, Bibhutibhusan, and Avaddhesh Narayan Singh (1962), *History of Hindu Mathematics: A Source Book*, single volume edition, Bombay: Asia Publishing House.

Dauben, Joseph W. (1998), "Ancient Chinese Mathematics: The 九章筭術 (*Jiu Zhang Suan Shu*) vs Euclid's *Elements*. Aspects of Proof and the Linguistic Limits of Knowledge," *International Journal of Engineering Science*, 36: 1339–59.

Dauben, Joseph W. (2007), "Chinese Mathematics," in Victor J. Katz, ed., *The Mathematics of Egypt, Mesopotamia, China, India, and Islam: A Sourcebook*, 187–384, Princeton, NJ: Princeton University Press.

Dauben, Joseph W. (2021), "Anachronism and Incommensurability: Words, Concepts, Contexts, and Intentions," in N. Guicciardini, ed., *Anachronisms in the History of Mathematics: Essays on the Historical Interpretation of Mathematical Texts*, Proceedings of the 2018 Bacon Conference, Division of the Arts and Humanities, Caltech, 307–57, Cambridge: Cambridge University Press.

De Young, Gregg (2005), "Diagrams in the Arabic Euclidean Tradition: A Preliminary Assessment," *Historia Mathematica*, 32 (2): 129–79.

Dhammaloka, Jambugahapitiye Thero (2019), "Śrīpati's Arithmetic in the *Siddhāntaśekhara* and *Gaṇitatilaka*: Edition, Translation and Mathematical and Historical Analysis," PhD diss., University of Canterbury, New Zealand.

Dharmachakra Translation Committee (2013), *Lalitavistara*. Available online: https://read.84000.co/translation/toh95.html?id=&part=UT22084-046-001-chapter-1 (accessed December 9, 2022).

Dold-Samplonius, Yvonne (1992), "Practical Arabic Mathematics: Measuring the Muqarnas by al-Kāshī," *Centaurus*, 35 (3): 193–242.

Dold-Samplonius, Yvonne, Joseph W. Dauben, Menso Folkerts, and Benno van Dalen, eds. (2002), *From China to Paris: 2000 Years of Mathematical Transmission*, Boethius, #46, Stuttgart: Franz Steiner.

Duits, Rembrandt (2005), "Celestial Transmissions: An Iconographical Classification of Constellation Cycles in Manuscripts (8th–15th Centuries)," *Scriptorium*, 59 (2): 147–202.

Eastwood, Bruce (2013), "Early-Medieval Cosmology, Astronomy, and Mathematics," in David C. Lindberg and Michael H. Shank, eds. *The Cambridge History of Science, vol. 2: Medieval Science*, 302–22, Cambridge: Cambridge University Press.

Englard, Yaffa (2003), "Mosaics as Midrash: The Zodiacs of The Ancient Synagogues and the Conflict between Judaism and Christianity," *Review of Rabbinic Judaism*, 6 (2): 189–214.

Evans, Gillian R. (1976), "The 'Sub-Euclidean' Geometry of the Earlier Middle Ages, Up to the Mid-Twelfth Century," *Archive for History of Exact Sciences*, 16 (2): 105–18.

Ferguson, James F. (1854), "A Calendar of the Contents of the Red Book of the Irish Exchequer," *Proceedings and Transactions of the Kilkenny and South-East of Ireland Archaeological Society*, 3 (1): 35–52.

Fine, Steven (2005), *Art and Judaism in the Greco-Roman World*, Cambridge: Cambridge University Press.

Goldstein, Bernard R. (1996), "Astronomy and Astrology in the Works of Abraham ibn Ezra," *Arabic Sciences and Philosophy*, 6 (1): 9–21.

Grant, Edward (1996), *The Foundations of Modern Science in the Middle Ages: Their Religious, Institutional, and Intellectual Contexts*, Cambridge: Cambridge University Press.

Greenwood, Kevin R. E. (2015), "The Meru Cosmos in Buddhist Art and Culture. Review of *Portrait of a Landscape: Depictions of the Meru Cosmos in Buddhist Art and Culture* by Eric Huntington," *Dissertation Reviews*, May 26.

Guicciardini, Niccolò, ed. (2021), *Anachronisms in the History of Mathematics: Essays on the Historical Interpretation of Mathematical Texts*, Proceedings of the 2018 Bacon Conference, Division of the Arts and Humanities, Caltech, Cambridge: Cambridge University Press.

Günther, Sebastian, ed. (2020), *Knowledge and Education in Classical Islam: Religious Learning between Continuity and Change*, 2 vols, Leiden: Brill.

Guo, Shuchun, ed. (2010), 中国科学技术史数学卷 *Zhongguo kexue jishushi—shuxue juan* (History of Ancient Chinese Science and Technology—Mathematics), Beijing: Science Press.

Gutas, Dimitri (1998), *Greek Thought, Arabic Culture: The Graeco-Arabic Translation Movement in Baghdad and Early 'Abbāsid Society (2nd–4th/8th–10th Centuries)*, London: Routledge.

Habib, Irfan (2008), *Technology in Medieval India: c. 650–1750*, Delhi: Tulika Books.

Hadley, John, and David Singmaster (1992), "Problems to Sharpen the Young," *Mathematical Gazette*, 76 (475): 102–26.

Halliwell, James O. (1841), *Rara Mathematica*, London: Maynard.

Harley, J. B., and David Woodward, eds. (1992), *The History of Cartography, vol. 2, book 1: Cartography in the Traditional Islamic and South Asian Societies*, Chicago, IL: University of Chicago Press.

Hayashi, Takao (1995), *The Bakhshālī Manuscript: An Ancient Indian Mathematical Treatise*, Groningen: Egbert Forsten.

Hayashi, Takao, Clemency Montelle, and K. Ramasubramanian, eds. (2019), *Bhāskara-prabhā*, Culture and History of Mathematics, #11, Delhi: Hindustan Book Agency.

Hill, Donald R. (1993), *Islamic Science and Engineering*, Edinburgh: Edinburgh University Press.

Hopkins, Jasper, trans. (2001), *Nicholas of Cusa's Complete Philosophical and Theological Treatises*, Minneapolis, MN: Arthur J. Banning Press.

Horiuchi, Annick (1994), *Les mathématiques japonaises à l'époque d'Edo*, Paris: J. Vrin.

Høyrup, Jens (2007), *Jacopo da Firenze's Tractatus Algorismi and Early Italian Abbacus Culture*, Science Networks, Historical Studies, #34, Basel: Birkhäuser.

Høyrup, Jens (2023), *The World of the Abbaco: Abbacus Mathematics Analyzed and Situated Historically Between Fibonacci and Stifel*, Basel: Birkhauser.

Huff, Toby E. (2017), *The Rise of Early Modern Science: Islam, China, and the West*, Cambridge: Cambridge University Press.

Huntington, Eric (2018), *Creating the Universe: Depictions of the Cosmos in Himalayan Buddhism*, Seattle, WA: University of Washington Press.

Ibn Khaldūn (2015), *The Muqaddimah: An Introduction to History*, ed. Nessim J. Dawood and trans. Franz Rosenthal, abridged edition, Bollingen Series, Princeton Classics, #13, Princeton, NJ: Princeton University Press.

İhsanoğlu, Ekmeleddin (2021), "Science in the Ottoman Empire," in *Studies on Ottoman Science and Culture*, 3–25, London: Routledge.

Islamic Scientific Manuscripts Initiative (n.d.), Available online: https://ismi.mpiwg-berlin.mpg.de/biographies-list (accessed December 9, 2022).
Jaki, Stanley L. (1974), *Science and Creation: From Eternal Cycles to an Oscillating Universe*, New York, NY: University Press of America (reprinted 1986, 1990).
Jeffreys, Elizabeth, John Haldon, and Robin Cormack, eds. (2008), *Oxford Handbook of Byzantine Studies*, Oxford: Oxford University Press.
Jiang, Xiaoyuan, ed. (2021), *The High Tide of Science and Technology Development in China*, Singapore: Springer Nature.
Katz, Victor J., ed. (2007), *The Mathematics of Egypt, Mesopotamia, China, India, and Islam: A Sourcebook*, Princeton, NJ: Princeton University Press.
Katz, Victor J., ed. (2016), *Sourcebook in the Mathematics of Medieval Europe and North Africa*, Princeton, NJ: Princeton University Press.
Katz, Victor J. (2017), "The Mathematical Cultures of Medieval Europe—Mathematics in Catholic Europe," *Convergence*. Available online: https://www.maa.org/press/periodicals/convergence/the-mathematical-cultures-of-medieval-europe-mathematics-in-catholic-europe (accessed December 9, 2022).
Keller, Agathe (2006), "Comment on a écrit les nombres dans le sous-continent indien, histories et enjeux," *Comptes rendus des séances de l'Académie des Inscriptions et Belles-Lettres*, 150 (4): 1879–96.
Kennedy, Edward S. (1956), "A Survey of Islamic Astronomical Tables," *Transactions of the American Philosophical Society*, 46 (2): 123–77.
Kennedy, Edward S. (1976), *Al-Bīrūnī's Exhaustive Treatise on Shadows*, 2 vols, Aleppo: Institute for the History of Arabic Science.
King, David A. (1983), "The Astronomy of the Mamluks," *Isis*, 74 (4): 531–55.
King, David A. (2007), "Ibn al-Shāṭir: ʿAlāʾ al-Dīn ʿAlī ibn Ibrāhīm," in Thomas Hockey, ed., *The Biographical Encyclopedia of Astronomers*, 569–70, New York, NY: Springer.
King, David A., and Julio Samsó (2001), "Astronomical Handbooks and Tables from the Islamic World (700–1900): An Interim Report," *Suhayl*, 2: 9–105.
Kobayashi, Haruo, and Mizue Kato, eds. (2010), *Transmission of Sciences: Greek, Syriac, Arabic and Latin*, Tokyo: Waseda Institute for Advanced Study.
Lam, Lay Yong (1997), "Zhang Qiujian Suanjing (The Mathematical Classic of Zhang Qiujian): An Overview," *Archive for History of Exact Sciences*, 50 (3/4): 201–40.
Lam, Lay Yong, and Ang Tian Se (2004), *Fleeting Footsteps: Tracing the Conception of Arithmetic and Algebra in Ancient China*, revised edition, Singapore: World Scientific.
Lewicka, Paulina (2011), *Food and Foodways of Medieval Cairenes: Aspects of Life in an Islamic Metropolis of the Eastern Mediterranean*, Leiden: Brill.
Lewis, Michael J. T. (2004), *Surveying Instruments of Greece and Rome*, Cambridge: Cambridge University Press.
Li, Liang (2016), "Arabic Astronomical Tables in China: Tabular Layout and its Implications for the Transmission and Use of the *Huihui lifa*," *East Asian Science, Technology and Medicine*, 44 (1): 21–68.
Li, Yan, and Du Shiran (1987), *Chinese Mathematics: A Concise History*, trans. John N. Crossley and Anthony W.-C. Lun, Oxford: Clarendon Press.
Li, Ye (Li Zhi) (1985), 敬齋古今黈. *Jingzhai gujin tou* (Thoughts of Jingzhai (Li Zhi) on [Things] Ancient and Modern), reprint, Taipei: Xinwenfeng Press.
Lim, Tina Su Lyn, and Donald B. Wagner (2017), *The Continuation of Ancient Mathematics: Wang Xiaotong's Jigu suanjing, Algebra and Geometry in 7th-Century China*, NIAS Reports, #51, Copenhagen: Nordic Institute of Asian Studies Press, University of Copenhagen.

Lindberg, David C. (1976), *Theories of Vision from al-Kindi to Kepler*, Chicago, IL: University of Chicago Press.

Lindberg, David C. (1992), *The Beginnings of Western Science: The European Scientific Tradition in Philosophical, Religious, and Institutional Context, 600 B.C. to A.D. 1450*, Chicago, IL: University of Chicago Press.

Lindberg, David C., and Michael H. Shank, eds. (2013), *The Cambridge History of Science, vol. 2: Medieval Science*, Cambridge: Cambridge University Press.

Lindblom, Jeannette (2019), "Women and Public Space: Social Codes and Female Presence in the Byzantine Urban Society of the 6th to the 8th Centuries," PhD diss., University of Helsinki.

Liu, Heping (2002), "'The Water Mill' and Northern Song Imperial Patronage of Art, Commerce, and Science," *Art Bulletin*, 84 (4): 566–95.

Lo, Andrew (2000), "The Game of Leaves: An Inquiry into the Origin of Chinese Playing Cards," *Bulletin of the School of Oriental and African Studies*, 63 (3): 389–406.

Lopez, Robert S., and Irving W. Raymond (1967), *Medieval Trade in the Mediterranean World: Illustrative Documents*, New York, NY: Norton.

Marmura, Michael E. (1968), "Causation in Islamic Thought," in Philip P. Wiener, ed., *Dictionary of the History of Ideas*, vol. 1, 286–9, New York, NY: Scribner's.

Martzloff, Jean-Claude (2006), *A History of Chinese Mathematics*, trans. Stephen S. Wilson, revised edition, Berlin: Springer.

Martzloff, Jean-Claude (2016), *Astronomy and Calendars: The Other Chinese Mathematics*, Berlin: Springer.

McEvilley, Thomas (2002), *The Shape of Ancient Thought: Comparative Studies in Greek and Indian Philosophies*, New York, NY: Allworth Press.

Merdinger, Charles J. (1954), "Surveying through the Ages, part I: Early Developments," *Military Engineer*, 46 (310): 99–102.

Misra, Anuj (2016), "The *Golādhyāya* of Nityānanda's *Sarvasiddhāntarāja*: An Examination of 'The Chapter on Spheres' in a Seventeenth Century Text of Mathematical Astronomy," PhD diss., University of Canterbury, New Zealand.

Montelle, Clemency (2011), *Chasing Shadows: Mathematics, Astronomy, and the Early History of Eclipse Reckoning*, Baltimore, MD: Johns Hopkins University Press.

Montelle, Clemency, and Kim Plofker (2018), *Sanskrit Astronomical Tables*, Cham: Springer.

Morel, Thomas (2020), "*De Re Geometrica*: Writing, Drawing, and Preaching Mathematics in Early Modern Mines," *Isis*, 111 (1): 22–45.

Morrow, Glenn R. (1970), *Proclus: A Commentary on the First Book of Euclid's Elements*, Princeton, NJ: Princeton University Press.

Needham, Joseph, with the cooperation of Wang Ling (1959), *Science and Civilisation in China, Vol. 3:Mathematics and the Sciences of the Heavens and the Earth*, Cambridge: Cambridge University Press.

Needham, Joseph, with the cooperation of Wang Ling and Kenneth Girdwood Robinson (1962), *Science and Civilisation in China, vol. 4: Physics and Physical Technology. Part I: Physics*, Cambridge: Cambridge University Press.

Needham, Joseph, with the cooperation of Wang Ling (1965), *Science and Civilisation in China, vol. 4: Physics and Physical Technology. Part II: Mechanical Engineering*, Cambridge: Cambridge University Press, 1965.

Needham, Joseph (1976), *Moulds of Understanding: Pattern of Natural Philosophy*, London: Allen & Unwin.

Needham, Joseph, Wang Ling, and Derek J. de Solla Price (1986), *Heavenly Clockwork: The Great Astronomical Clocks of Medieval China*, 2nd edn, with suppl. by John H. Combridge, Cambridge: Cambridge University Press.

North, John D. (2008), *Cosmos: An Illustrated History of Astronomy and Cosmology*, Chicago, IL: University of Chicago Press.

Orme, Nicholas (2006), *Medieval Schools: From Roman Britain to Renaissance England*, New Haven, CT: Yale University Press.

Paternicò, Luisa M. (2018), "Jacob Golius and Martino Martini: The Enlightening Encounter and the Additamentum," in Chen Tsung-ming, ed., *Catholicism's Encounters with China: 17th to 20th Century*, Leuven Chinese Studies, #39, 185–205, Leuven: Ferdinand Verbiest Institute.

Petrocchi, Alessandra (2017), "The *Gaṇitatilaka* and its Commentary by Siṃhatilakasūri: An Annotated Translation and Study," PhD diss., University of Cambridge.

Pingree, David (1968), "The Fragments of the Works of Ya'qūb Ibn Ṭāriq," *Journal of Near Eastern Studies*, 27 (2): 97–125.

Pingree, David (1981), *Jyotiḥśāstra*, Wiesbaden: Harrassowitz.

Plofker, Kim (2007), "Mathematics in India," in Victor J. Katz, ed., *The Mathematics of Egypt, Mesopotamia, China, India, and Islam: A Sourcebook*, 385–514, Princeton, NJ: Princeton University Press.

Plofker, Kim (2009), *Mathematics in India*, Princeton, NJ: Princeton University Press.

Plofker, Kim, and Toke L. Knudsen (2011), "Calendars in India," in John M. Steele, ed., *Calendars and Years II: Astronomy and Time in the Ancient and Medieval World*, 53–68, Oxford: Oxbow Books.

Ramakalyani, K. (2016), "A Critical Study of the Commentary *Buddhivilasini* of Ganesa Saivajna on *Lilavati* of Bhaskāra," PhD diss., Kuppuswami Sastri Research Institute, Chennai. Available online: http://hdl.handle.net/10603/295801 (accessed December 9, 2022).

Ramasubramanian, K., M. D. Srinivas, and M. S. Sriram (1994), "Modification of the Earlier Indian Planetary Theory by the Kerala Astronomers (*c.* 1500 AD) and the Implied Heliocentric Picture of Planetary Motion," *Current Science*, 66 (10): 784–90.

Rangacarya, M. (1912), *Ganitasarasangraha of Mahavira*, Madras: Government Press.

Reding, Jean-Paul (2004), *Comparative Essays in Early Greek and Chinese Rational Thinking*, Aldershot: Ashgate.

Rosen, Frederic (1831), *The Algebra of Mohammad ben Musa*, London: Oriental Translation Fund.

Sabra, Abdelhamid I. (2003), "Ibn al-Haytham: Brief Life of an Arab Mathematician: Died *circa* 1040," *Harvard Magazine*, September/October. Available online: https://harvardmagazine.com/2003/09/ibn-al-haytham-html (accessed December 9, 2022).

Sachau, Edward C. (1910), *Alberuni's India*, London: Kegan, Paul, Trench, Trübner.

Saidan, A. S. (2011), *The Arithmetic of Al-Uqlīdisī*, Heidelberg: Springer.

Saito, Ken (2018), "Diagrams in Arithmetical Books of Euclid's *Elements*," *SCIAMVS*, 19: 103–66.

Saito, Ken, and Nathan Sidoli (2012), "Diagrams and Arguments in Ancient Greek Mathematics: Lessons Drawn from Comparisons of the Manuscript Diagrams with those in Modern Critical Editions," in Karine Chemla, ed., *The History of Mathematical Proof in Ancient Traditions*, 135–62, Cambridge: Cambridge University Press.

Saliba, George (1994), *A History of Arabic Astronomy: Planetary Theories during the Golden Age of Islam*, New York, NY: New York University Press.

Sarhangi, Reza, and Slavik Jablan (2006), "Elementary Constructions of Persian Mosaics," *Math Horizons*, 14 (1): 10–13.

Sarma, Sreeramula Rajeswara (2008), *The Archaic and the Exotic: Studies in the History of Indian Astronomical Instruments*, New Delhi: Manohar.

Sarma, Sreeramula Rajeswara (2018), *A Descriptive Catalogue of Indian Astronomical Instruments*. Available online: https://www.srsarma.in/catalogue.php (accessed December 9, 2022).

Savage-Smith, Emilie (2013), "The Most Authoritative Copy of 'Abd al-Raḥmān al-Ṣūfī's Tenth-Century Guide to the Constellations," in Sheila Blair and Jonathan Bloom, eds. *God is Beautiful; He Loves Beauty: The Object in Islamic Art and Culture*, 122–55, New Haven, CT: Yale University Press.

Sayili, Aydin (1956), "'Alā al-Dīn al-Manṣūr's Poems on the Istanbul Observatory," *Belleten*, 20 (79): 429–84.

Schmidl, Petra G. (1997–8), "Two Early Arabic Sources on the Magnetic Compass," *Journal of Arabic and Islamic Studies*, 1: 81–132.

Schwartzberg, Joseph E. (1992), "Introduction to South Asian Cartography," in J. B. Harley and David Woodward, eds. *The History of Cartography, vol. 2, book 1: Cartography in the Traditional Islamic and South Asian Societies*, 295–331, Chicago, IL: University of Chicago Press.

Sebregondi, Giulia, and Richard Schofield (2016), "First Principles: Gabriele Stornaloco and Milan Cathedral," *Architectural History*, 59: 63–122.

Selin, Helaine, ed. (2008), *Encyclopaedia of the History of Science, Technology, and Medicine in Non-Western Cultures*, 2nd edn, Berlin: Springer.

Shatzmiller, Maya (1988), "Aspects of Women's Participation in the Economic Life of Later Medieval Islam: Occupations and Mentalities," *Arabica*, 35 (1): 36–58.

Shen, Kangshen (1999), *The Nine Chapters on the Mathematical Art: Companion and Commentary*, trans. John N. Crossley and Anthony W.-C. Lun, Oxford: Oxford University Press.

Shi, Yunli (2014), "Islamic Astronomy in the Service of Yuan and Ming Monarchs," *Suhayl*, 13: 41–61.

Shi, Yunli (2021), "Astronomy and Diplomacy: Official Contacts in Astronomy between China and Korea," in Xiaoyuan Jiang, ed., *The High Tide of Science and Technology Development in China*, 361–95, Singapore: Springer Nature.

Shukla, Kripa Shankar (1976), *Āryabhaṭīya of Āryabhaṭa with the Commentary of Bhāskara I and Someśvara*, New Delhi: Indian National Science Academy.

Sigler, Laurence E. (2002), *Fibonacci's Liber Abaci*, New York, NY: Springer.

Singh, Upinder (2008), *A History of Ancient and Early Medieval India: From the Stone Age to the 12th Century*, Delhi: Pearson Education India.

Sivin, Nathan (2009), *Granting the Seasons: The Chinese Astronomical Reform of 1280, with a Study of its Many Dimensions and a Translation of its Records*, New York, NY: Springer Science+Business Media.

Smith, David E., and Louis Charles Karpinksi (1911), *The Hindu-Arabic Numerals*, Boston, MA: Ginn.

Sparavigna, Amelia Carolina (2017), "Magnetic Compasses and Chinese Architectures," *Philica*, art. 1094. Available online: https://doi.org/10.5281/zenodo.4984403 (accessed December 9, 2022).

Spufford, Peter (1989), *Money and its Use in Medieval Europe*, Cambridge: Cambridge University Press.

Srinivas, M. D. (2022), "Emergence of a New Era in the History of Indian Mathematics: Historiography of the Kerala School of Astronomy and Mathematics, 1820–2021: Some Highlights," *Bhāvanā*, 6 (1). Available online: https://bhavana.org.in/emergence-of-a-new-era-in-the-history-of-indian-mathematics/ (accessed December 9, 2022).

Srinivasan, Saradha (1979), *Mensuration in Ancient India*, Delhi: Ajanta Publications.

Stahl, William H. (1977), *Martianus Capella and the Seven Liberal Arts*, vol. 2, New York, NY: Columbia University Press.

Steele, Robert (1922), *The Earliest Arithmetics in English*, London: Oxford University Press.

Steele, John M., ed. (2011), *Calendars and Years II: Astronomy and Time in the Ancient and Medieval World*, Oxford: Oxbow Books.

Struik, Dirk J. (1967), *A Concise History of Mathematics*, 3rd revised edition, New York, NY: Dover.

Swetz, Frank (1987), *Capitalism and Arithmetic*, La Salle, IL: Open Court Publishing.

Tihon, Anne (2008), "Numeracy and Science,' in Elizabeth Jeffreys, John Haldon, and Robin Cormack, eds. *Oxford Handbook of Byzantine Studies*, 803–19, Oxford: Oxford University Press.

Trimble, Virginia, Thomas R. Williams, Katherine Bracher, Richard A. Jarrell, Jordan D. Marché II, JoAnn Palmeri, and Daniel W. E. Green, eds. (2007), *Biographical Encyclopedia of Astronomers*, New York, NY: Springer Science+Business Media.

Tsung-ming, Chen, ed. (2018), *Catholicism's Encounters with China: 17th to 20th Century*, Leuven Chinese Studies, #39, Leuven: Ferdinand Verbiest Institute.

Vogel, Kurt (2002), "A Surveying Problem Travels from China to Paris," in Yvonne Dold-Samplonius, Joseph W. Dauben, Menso Folkerts, and Benno van Dalen, eds. *From China to Paris: 2000 Years of Mathematical Transmission*, Boethius, #46, 1–7, Stuttgart: Franz Steiner.

Wallis, Faith (1999), *Bede: The Reckoning of Time*, Liverpool: Liverpool University Press.

Wiener, Philip P., ed. (1968), *Dictionary of the History of Ideas*, vol. 1, New York, NY: Scribner's.

Williams, Jack (1995), "Mathematics and the Alloying of Coinage 1202–1700," *Annals of Science*, 52 (3): Part I, 213–34; Part II, 235–63.

Woodhead, Christine (1983), "An Experiment in Official Historiography: The Post of Şehnāmecī in the Ottoman Empire, *c.* 1555–1605," *Wiener Zeitschrift für die Kunde des Morgenlandes*, 75: 157–82.

Yabuuti, Kiyosi (1979), "Researches on the *Chiu-chih li*—Indian Astronomy under the Tang Dynasty," *Acta Asiatica*, 36: 7–48.

Yano, Michio (1986), "Knowledge of Astronomy in Sanskrit Texts of Architecture (Orientation Methods in the *Iśānaśivagurudevapaddhati*)," *Indo-Iranian Journal*, 29 (1): 17–29.

INDEX

Entries in italics indicate pages with illustrations.